PROGRESS IN PLANNING

Editors

D. Diamond & B. H. Massam

PERGAMON

Aims and Scope

To provide a convenient and rapid outlet for extended papers drawn from recent social science research of importance for the practice of urban and regional planning and environmental management. New and innovatory aspects of planning are emphasized: the contributors are international in character and material is drawn from the entire range of social sciences throughout the world.

The range of disciplinary interests covered include: sociology (especially urban sociology), demography, development planning, environmental studies, economic, social, urban, rural and population geography, civic design, transportation planning, political science and public administration, social administration and urban and regional economics. The methodology of planning, the philosophy and purpose of planning and its relationship to long-term forecasting, the management of information systems and social theory are also included.

Progress in Planning covers the following subject categories:

Transport Planning
Planning Theory and Method
Regional and Economic Planning
Third World Planning
Industry and Employment
Physical, Urban and Environmental Planning

∞™ The paper used in this publication meets the minimum requirements of American National Standard for Information Sciences—Permanence of Paper for Printed Library Materials, ANSI Z39.48-1984.

Transferred to digital printing 2006

Printed and bound by Antony Rowe Ltd, Eastbourne

Political economy and urban planning: a comparative study of Hong Kong, Singapore and Taiwan

MEE KAM NG

Centre of Urban Planning and Environmental Management,
The University of Hong Kong, Pokfulam Road, Hong Kong

PERGAMON

Biography

Mee Kam Ng is an assistant professor at the Centre of Urban Planning and Environmental Management, the University of Hong Kong. She obtained her Ph.D. in 1992 from the University of California, Los Angeles under the supervision of Professor John Friedmann. Her research focuses on urban and regional development and planning in Pacific Asia. Current research projects include: a comparative study of the land use planning systems in Hong Kong and selected cities in Guangdong province, China and advocacy planning in Hong Kong. Recent published works include: Reclamation: An Urban Development Strategy under Fire, *Land Use Policy* **14**(1), 5–23, 1997 (with A. Cook); A Critique of the City Planning Act of the People's Republic of China: A Western Perspective, *Third World Planning Review* **17**(3), 279–293 (with F. Wu); and Urban and Regional Planning, In *From Colony to SAR: Hong Kong's Challenges Ahead*, eds Cheng, J. Y. S. and Lo, S. Chinese University Press, Hong Kong, pp. 227–260 (1995). She is a full member of the Royal Town Planning Institute and the Hong Kong Institute of Planners. Her published works received a Certificate of Merit and a Silver Medal in the Hong Kong Institute of Planners Award 1994 and 1995 respectively.

Contents

Pergamon

Progress in Planning Vol. 9, pp. 1–90, 1999

Printed in Great Britain
0305-9006/99/$ - see front matter

PII: S0305-9006(98)00027-0

Abstract

This study compares urban planning mechanisms that operate within Hong Kong, Singapore and Taiwan. The political economy of Hong Kong is in a state of flux. While the power of the government and the corporate interests remain largely intact, they are challenged by pro-China interests and a democratizing civil society. The land use planning system reflects this power contest. In face of both strong resistance from the development industry and China's eagerness to perpetuate a market-led society in post-1997 Hong Kong, the outcome of the power contest remains uncertain. The state-centred political economy of Singapore has bred a top–down land use planning system centrally controlled by the government. Not only has the government dominated the plan making process, the legislation has entrusted the public sector to scrutinize and guide private development through a discretionary development control system. The government is able to mobilize resources to implement plans with the tacit consent of a regulated and meritocracy-based society. In contrast, Taiwan's multi-layered government structure and its complicated relationships with business interests ('gold-power' alliance) within a cultural milieu with scant respect for rules and regulations, have produced a complex and yet loosely coordinated land use planning system. The Urban Planning Law emphasizes plan making and implementation by the public sector at the district level, but local authorities lack resources to implement plans. The regulatory development control system, to a large extent, is abused by the community and land use zoning is not taken seriously in general. As plan amendments can be made behind closed doors, the 'gold-power' alliance has tried to rezone land for speculative purposes. The impact of the recent democratization process on the society's legal attitudes and the roles of planning in urban governance in Taiwan remains to be seen.

Acknowledgements

I would like to thank the Committee for Research and Conference Grants, The University of Hong Kong for a research grant to carry out this project. The assistance of my friends in Taiwan and Singapore is deeply appreciated. I would also like to thank my colleague, A. Cook, for her valuable comments on an earlier draft of the paper. While my former professors at UCLA may not be aware of this research project, the approach adopted in this comparative study is deeply influenced by the knowledge that I gained, synthesized and internalized from them. All the faults in this paper, of course, remain my sole responsibility.

CHAPTER 1

Introduction

Most of the comparative studies on urban planning mechanisms are about Western countries* (Booth, 1996; Cullingworth, 1993; Faludi and Hamnett, 1975; Johnson, 1984; Strong, 1971; Thomas, 1983; White, 1979). In the Asian context, there are many comparative research studies on the economic miracles and the roles of the states in development (Appelbaum and Henderson, 1992; Deyo, 1987; Douglass, 1994; Geiger, 1973; Hughes, 1988; International Labour Organization, 1981; Lee, K., 1993; MacIntyre, 1994; Perkins, 1994; Wade, 1990; World Bank, 1993). However, there seems to be no comparative work on urban planning mechanisms in the 'little dragons' in Asia.† In order to fill this gap, this study will compare urban planning mechanisms in Hong Kong, Singapore and Taiwan. It is hoped that, through a comparative study of these three political economies, useful lessons can be learnt by other Asian countries.

The following chapter discusses the study objectives and some methodological issues involved in comparative studies. An outline of the theoretical framework adopted in this study is presented. It is argued that the political economy of a country will provide the key to understanding the nature and content of the urban planning mechanisms. Hence, in Chapter 3, the historical development and political economies of the three cases are examined. This establishes the background for appreciating the institutionalization, operation, and success or failure of the urban planning mechanisms in the three economies. Chapters 4–7 compare the land use planning mechanisms in Hong Kong, Singapore and Taiwan: the planning administration; hierarchy of plans; land use planning legislation and the plan making process; and development control and plan implementation. Chapter 8 concludes the monograph and suggests a research agenda for comparative planning studies in the Asian context.

---

*A few exceptions can be found: Khakee (1992), Ministry of Construction and Transport, South Korea (1996), Ng and Xu (1996).

†There is, however, research on urban planning in the Asian context such as Heikkila and Griffin (1995).

CHAPTER 2

Theoretical framework

2.1. STUDY OBJECTIVES AND METHODOLOGICAL ISSUES

This exploratory study aims to understand the urban planning mechanisms in Hong Kong, Singapore and Taiwan, three of the 'little dragons' in Asia.* While it is widely acknowledged that the three dragons are successful newly industrialized economies, there is little research done to ascertain the role of urban planning in the miraculous economic development, or how land is planned and restructured to accommodate economic growth.† It is hoped that this study is a modest step towards this end.

Sewell argues that 'the comparative method is essentially a tool for dealing with problems of explanation' (Sewell, 1967, p. 208). Masser (1984, p. 148) also suggests that '[c]ross national comparative planning studies are valuable for two reasons: as a means of improving planning practice through the interpretation and transfer of experience from one country to another; and as a way of developing planning theory by transcending national cultural boundaries.' Cullingworth (1993) argues that policies are cultural products of history, time and space and therefore are rarely exportable. However, 'comparisons prompt new questions and, in contemplating these, new policy initiatives may emerge, molded to fit in a different political terrain' (p. 177). As asserted by Wolman (1985, p. 116), 'comparative analysis···can both broaden the sense of the possible and provide a framework for better understanding of our behavior. The process of comparative analysis should lead to questioning what frequently is simply assumed without question'. While this study may not suggest new policy options in the field of urban planning, it is hoped that the political economies under study and other Asian countries can reflect upon their own experience and consider what alternatives are desirable or feasible.

It is a daunting task to put forward an explanatory theoretical framework for describing, comparing and analyzing the three cases. This study attempts to develop a tentative theoretical framework for understanding the institutionalization and operation of the urban planning mechanisms within a wider socio-economic and political

*The original idea was to undertake a comparative study on the four little dragons including South Korea. However, for various reasons, the author could not obtain information on South Korea until recently and so South Korea is excluded in this comparative study.

†General factors often quoted for the success of the Asian dragons are: historical legacies such as colonialism, geopolitics, world trade expansion; the rise of the new international division of labour as a result of the transaction revolution; export-led industrialization; the role of foreign capital; culture (particularly Confucianism); the existence of free markets; the developmental states, etc. (Appelbaum and Henderson, 1992; Douglass, 1994; International Labour Organization, 1981).

context. The theoretical framework developed is not only concerned with the urban planning mechanism but also with the context. The approach of this study is therefore different from Khakee's 'pragmatic' approach (Khakee, 1996, p. 95) which 'does not delve too deeply into the context'. According to Przeworski and Teune, comparative studies are those which examine and theorize the influence of larger systems (the setting or contexts) upon the characteristics (relationships) of units within them (Przeworski and Teune, 1970, pp. 70, 23). As '[p]lanning is as much an integral part of a particular political system and institutional fabric in a given country as is the specific process which it is attempting to guide· · ·the researchers must expand their study horizons beyond the narrow confines of planning practice to encompass the operation of the political system and policy-making environment in which that practice has evolved' (Bourne, 1984, p. 150). This point is repeated by Cullingworth (1993, p. 178), 'land use planning· · ·cannot be understood except through an appreciation of broader social, economic, political, and historical factors'.

The theoretical framework examined below is an attempt to understand the operations of the urban planning mechanisms within a wider socio-economic, political and spatial context. The units to be analyzed and compared are the urban planning mechanisms within the three specific political economies.

2.2. FROM POLITICAL ECONOMY TO URBAN PLANNING MECHANISMS

The political economy approach is adopted to construct the theoretical framework. Given the intricate differences of the three political economies (Douglass, 1994), it is difficult, if not impossible, to adopt Hobbs' approach in examining British town planning which, he argues, has 'originated and evolved in response to dysfunctions in the process of capital accumulation' (Hobbs, 1992, p. 193). His basic thesis is that 'the uneven temporal and spatial processes of economic change have structured the operation and evolution of post-war British town planning' (Hobbs, 1992, p. 191). While economic forces are important in shaping urban planning mechanisms in the three cases studied, it will be argued that even their economic development has to be understood in a wider context.

One major component of the tentative theoretical framework is an analysis of the power relationship among key actors in a political economy. Although the difficulties of transforming structural relationships are fully appreciated, this study adopts a dynamic perspective in understanding the power relationship among different actors. Amirahmadi and Gladstone's (Amirahmadi and Gladstone, 1996) dynamic theory of the state and civil society, which argues that the relative strength of the state and civil society would change over the course of the development process, is particularly relevant. As can be seen in Fig. 1, the tentative theoretical framework depicts a dynamic relationship between the political economy and urban planning mechanisms within a specific spatial-temporal cultural entity which is subject to external influences.

Since Hong Kong, Singapore and Taiwan are export-oriented economies which were colonies in recent decades, global–local nexus and legacies of world history

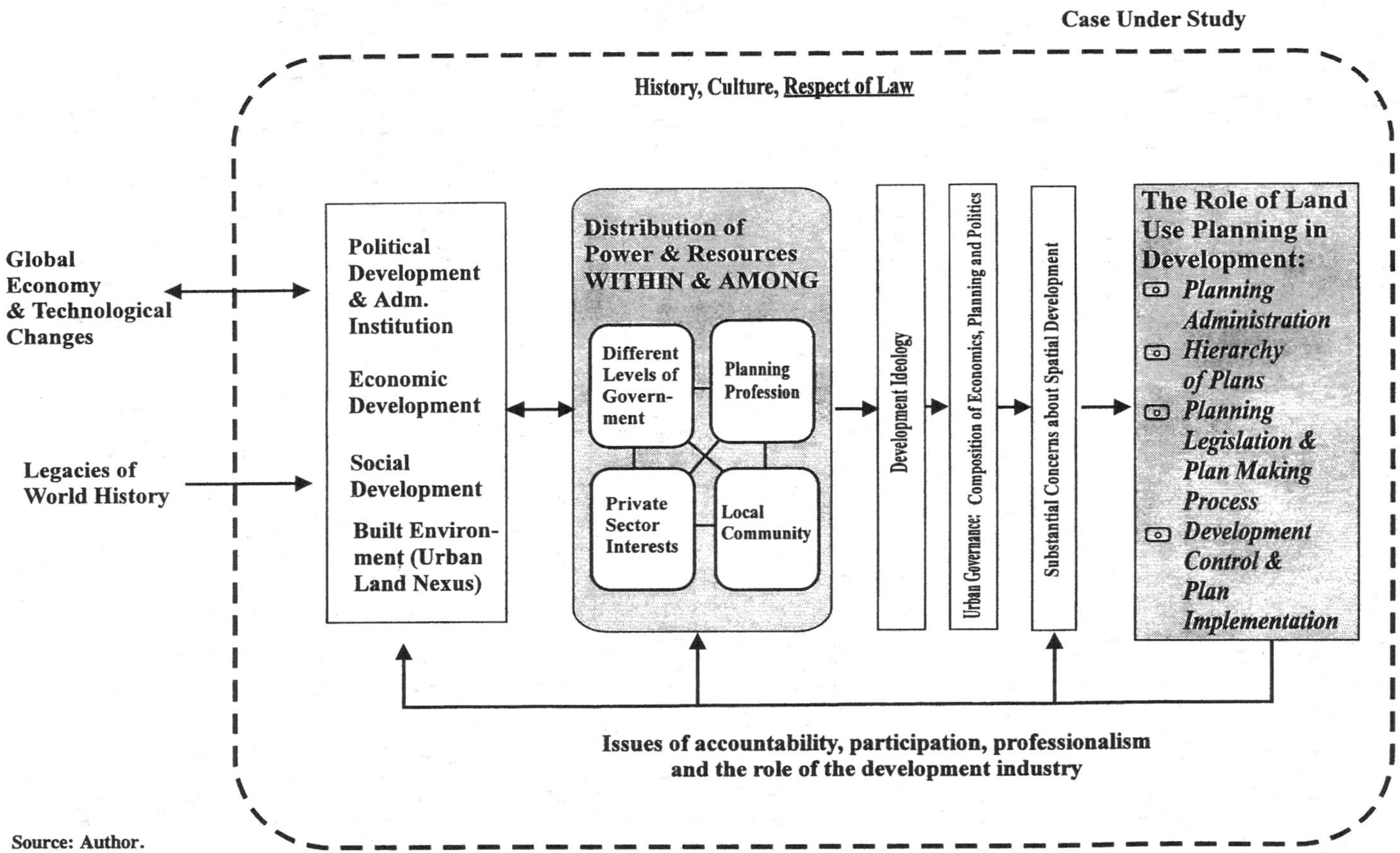

FIG. 1. A tentative theoretical framework. (Source: Author)

have interacted with domestic histories and culture to shape their socio-economic, political and spatial development. Yet in order to understand why a particular society embraces a specific development ideology, it is essential to untangle the power relationship among key actors in the development process. Four major groups or actors can be identified: the government, the planning professionals, private sector interests (especially those related to property development related industries) and local community groups. The role of the government in the political system, the development ideology advocated by the state and whether it can command resources in urban management, will influence how the urban planning mechanisms are established and implemented. The role of the local and overseas private sector interests in economic development, the cultural milieu, and the relative powers of the local community, the government and private sector will also determine the institutionalization and effectiveness of urban planning mechanisms. These in turn will determine a society's commitment to urban planning, and the professional and social status of urban planners.

In a more general sense, the power relationship in a society will define the relative importance of three major means of urban governance: economics, planning and politics. While the specific contents of these three terms vary in different spatial-temporal contexts, the substantial concerns about spatial development and the resultant urban forms of a place will reflect the balance of these three forces. In other words, the political economy of a place will determine the role of urban planning in its development trajectory.

However, it should be noted that while various urban planning mechanisms and the accompanying institutional set-up can be derived from the wider socio-economic and political context, the administration of these institutions and mechanisms may in turn influence the power relationship. If properly established, urban planning mechanisms can be used to maximize the contribution of professionals, ensure accountability of the government, enhance citizen participation, and encourage the development industries to be 'community builders'. However, as can be observed from the case studies, the social perception of rules and regulations basically determines the relevance of urban planning mechanisms in facilitating and controlling development. Unless state intervention and control in the built environment are taken seriously by a society, the urban planning system adopted will not be an effective one.

If urban planning rules and regulations are respected in a society, then many theoretical issues pertinent to Western planning systems will become relevant: accountability of the government, openness of the urban planning mechanisms, citizen participation, professional ethics and the question of who gets what, where, when and how in an urban planning process.

In the following chapter, background information of the three cases under study is examined. Some comparative statistics will be analyzed before the histories of the three political economies are discussed.

CHAPTER 3

Political economies and urban planning mechanisms in Hong Kong, Singapore and Taiwan

3.1. INTRODUCTION

Table 1 compares some key economic statistics for the three countries of the case study. In terms of land area, Singapore is the smallest (641.6 km^2) and Taiwan is about 36 times the size of Hong Kong (1092 km^2). While half of the land area in Singapore is urbanized, less than 15% of the land area in Hong Kong and Taiwan is urbanized. The population of Hong Kong (6.3 million) is double that of Singapore (about 3 million) and the population in Taiwan is 21.1 million. Population density is highest in Hong Kong (5890/km^2) and lowest in Taiwan (577/km^2) with Singapore in the middle (4612/km^2). However, in the city of Taipei, the capital of Taiwan, the population is 9762/km^2. Table 1 also illustrates the rapid economic growth of the three case countries in the past three decades. While industries are significant contributors to the Gross Domestic Product (GDP) in Singapore and Taiwan, industry in Hong Kong has declined since the 1990s. In 1995, the contribution of manufacturing industries to GDP in Hong Kong had dropped to below 20%.

Colonization has been part of the history of the three economies. Hong Kong was a British colony from 1842 to 1997. However, Hong Kong did not start its economic take-off until the 1950s, when mainland capitalists and refugees fled to Hong Kong to escape the Communist rule after the establishment of the People's Republic of China in 1949. In July 1997, Hong Kong became a Special Administrative Region in China under the 'One Country, Two Systems' concept agreed to in the 1984 Sino–British Joint Declaration. Singapore was also a British colony from 1824 to 1963. In 1959 Singapore attained internal self-government and in 1963 the tiny independent city-state joined the Federation of Malaysia. However, the merger was short-lived and in 1965 Singapore became an independent city-state. Taiwan was a Japanese colony before 1945 (the end of the Second World War) for 50 years. To many native Taiwanese, the post-1945 rule of the mainland Nationalist Government is another era of 'colonization'.

One may argue that these historical legacies have significantly influenced the three political economies which are examined in further detail below.

Table 1. Selected statistics on Hong Kong, Singapore and Taiwan

	Hong Kong	Singapore	Taiwan
Land area (km^2)	1092.0[a]	641.6[b]	36 000.0[c]
Urban land (%)	15.6[a]	49.7[b]	12.2[d]
Population (1995)	6 307 900.0[a]	2 986 500.0[e]	21 126 000.0[f]
Urban population (%)	94.0[g]	100.0[g]	77.3[d]
Population density 1995 (/km^2)	5890[a]	4612[e]	577[c]
Per capita real GDP[h] (in 1989, US$)			
1970	3053.0	1583.0	1110.0
1985	6722.0	6055.0	3094.0
2000	13 975.0	12 580.0	6995.0
Per capita GDP at current market price (US$) in 1994	21 417.0[i]	22 934.0[e]	11 604.0[f]
GDP at current price (million US$)			
1965	1783[i]	1000[h]	2811[f]
1975	6315[i]	4227[j]	15 429[f]
1985	34 828[i]	17 530[k]	63 097[f]
1994	129 819[l]	151 838[e]	244 169[f]
Average GDP annual growth rates (%)			
1965–1980[k]	8.5	9.3	10.2
1960–1970[m]	10.0	9.6	9.3
1970–1980[m]	9.0	9.6	9.8
1980–1985[m]	6.7	6.8	6.3
1980–1992[n]	6.7	10.6	6.7
Sectoral share of GDP (1995) (%)[o]			
Agriculture	0.2	0.2	3.1
Industry	18.1	42.7	38.9
Services	81.7	57.1	58.0

Sources: [a]: Government Information Services (1996, pp. 395, 468). [b]: Singapore Department of Statistics (1993). [c]: CEPD (1992). [d]: The figure for urban area refers to planned urban area. The figure for urban population refers to population in planned urban areas CEPD (1993, p. 2). [e]: Figure from the web site: http://www.singstat.gov.sg/KEYSTAT/POP.html on 4 January 1997. Assuming S$1 = US$0.712758. [f]: 1994 figure from: gopher://sun.cepd.gov.tw:70/00/databook/1-1t on 4 January 1997. GNP figure instead of GDP. [g]: Yeung (1993, Table 1.1, p. 7). [h]: Shinohara (1989, Table 1, pp. 7. [i]: Figure from: Hong Kong Census and Statistics Department (1994) (assuming an exchange rate of HK$7.8 = US$1). [j]: Singapore Department of Statistics (1983, pp. 7, 57) (assuming exchange rate of S$1 = US$0.338). [k]: Lo *et al.* (1989, p. 109. [l]: Figure from: http://www.info.gov.hk/censtatd/hkstat on 10 January 1997 (assuming exchange rate HK$7.8 = US$1). [m]: Akrasanee and Chirathivat (1989, p. 402). [n]: Far Eastern Economic Review (1995, p. 14). [o]: ADB (1996, p. 228).

3.2. THE POLITICAL ECONOMY OF HONG KONG: AN UNBALANCED POWER CONTEST

Hong Kong's manufacturing power grew in the 1950s, as a result of a 'transferred industrialization' process initiated by mainland capitalists who fled to Hong Kong on the establishment of the communist People's Republic of China in 1949. Given Hong Kong's history as a trading port, manufactured products easily found ways to be sold in the world market. Hence, unlike other Asian newly industrialized economies, Hong Kong did not go through an import-substitution industrialization period in its economic history. The colonial government had practised a 'laissez-faire' industrial development policy. As the largest landlord in the territory, the government has, since the inception of the colony, relied on land sales to boost the public coffers. One may argue that the government has a larger stake in the development-related industries (such as property and financial sectors) rather than the manufacturing ones.

The 155 years of British rule has changed quite fundamentally the originally Chinese value-laden political culture of the local residents. People in Hong Kong have high respect for laws, rules and regulations. Although democracy was not introduced in the local political scene until recently, since the late 1970s, people in Hong Kong have always enjoyed liberty, individual human rights and freedom of speech.

The signing of the Sino–British Joint Declaration in 1984, which marked the beginning of the transition period, could be seen as a watershed in Hong Kong's socio-economic and political development. Before the transition period, it could be argued that Hong Kong had an 'inter-corporate and British-related 'club' politics'. The executive-led 'administrative no-party' state was heavily influenced by the business community. Davies in 1975 analyzed the composition of various government advisory boards, committees and the three councils (the law-making Legislative Council, the most important policy taking Executive Council, and the Urban Council in charge of municipal services) and discovered that the business interests had an overwhelming voice in the consultation machinery (about 70% of the total membership) (Davies, 1977, p. 51). This bureaucratic-cum-corporatist state was seldom challenged by a politically apathetic refugee society, the prime concern of which was to earn a decent living. The export-oriented economy thrived in an era of expanding world trade in the 1960s. During those days, Hong Kong's society was relatively simple with the interests of government and the private sector dominating those of the community (including the professionals).

The socio-economic and political transition since the early 1980s has highlighted an increasingly complex relationship between the maturing and restructuring economy and a democratizing polity which reverts to Chinese rule in mid-1997. As the biggest landlord in the territory, the government still retains a strong position in commanding resources to direct spatial development in the territory. However, at the political front, the government not only faces the challenge of being a 'Special Administrative Region' (SAR) under Beijing's rule, it also needs to cope with demands raised by the Western educated, liberal-minded offspring of the refugee generation. In other words, with the end of colonial rule, the 'relative autonomy' of the government (from 'high' and local politics) is gone. While the Chinese government has tried to protect the business interests in running the SAR,* the democratization process in the last years of British colonial rule has nurtured the growth of a fledgling political community. The maturing political economy has led to the rising status of professional groups whose knowledge and expertise are required to meet challenges imposed by a restructuring economy during a period of political transition. Figure 2 maps out the intricate relationship between the various actors in a wider context.

While politics has become part of the life of Hong Kongers, the development ideology remains very much economic growth oriented. The invisible hand, the market, will still be a major means of urban governance in the SAR. The government has adopted a 'minimum intervention and maximum support' industrial policy in recent

---

*This could be seen in the selection of the Chief Executive and the provisional legislature of the SAR. The Selection Committee which was responsible for selecting the Chief Executive and the provisional legislature was dominated by pro-China groups. Tung Chee-hwa, the first Chief Executive of the SAR, was formerly a shipping magnate having close business interests with local and Chinese interests. Some of the prominent pro-China losers in the 1995 direct election of the Legislative Council were selected into the provisional legislature in 1996.

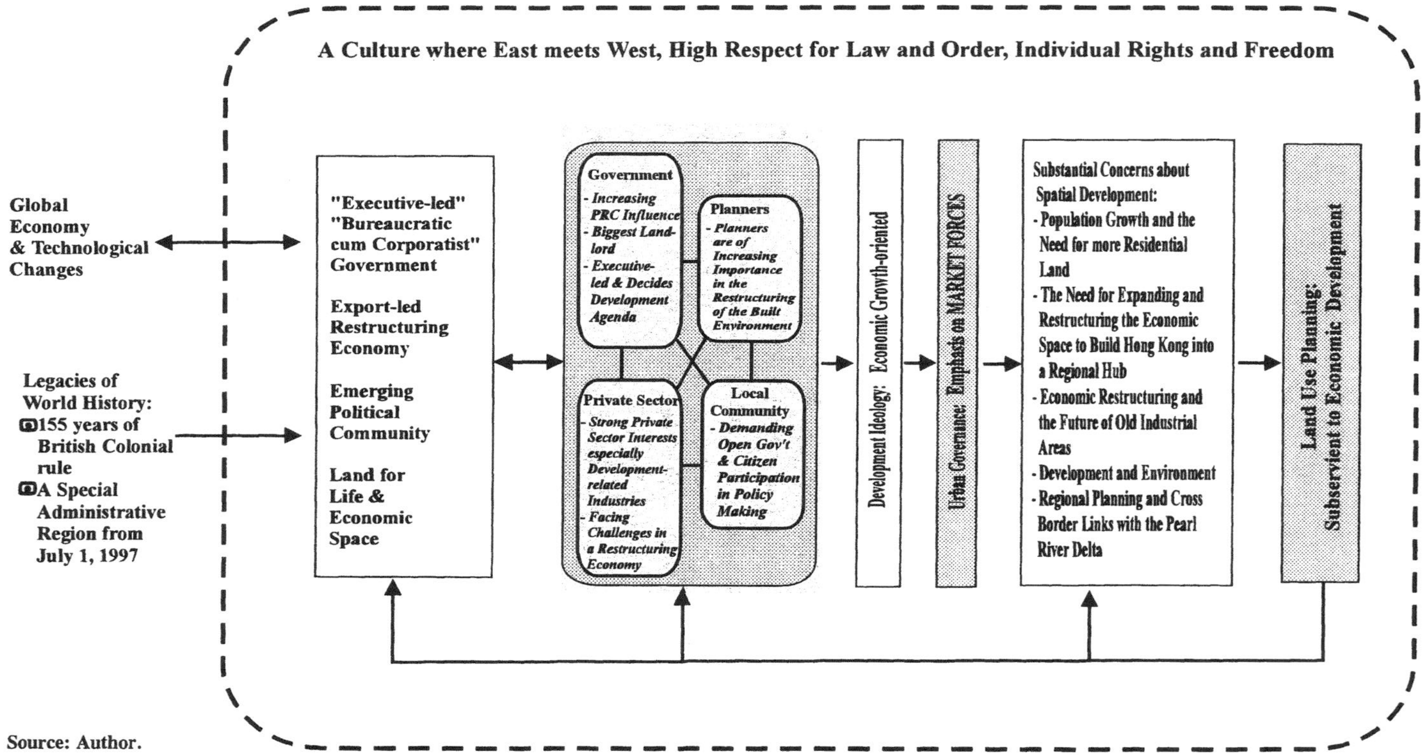

FIG. 2. Political economy and urban planning: the case of Hong Kong. (Source: Author)

years. It has maintained strong control over land supply and development—a major source of government revenue. In fact, as argued by Cuthbert (1985, p. 154), 'the monopoly ownership of all land by the State combined with the principle of sale by auction to the highest bidder set the context for super-profits on land both by the Government and finance capital generally'. It has also been argued that the government has deliberately pursued a reclamation-led urban development strategy for land supply to boost the public coffers (Ng and Cook, 1997). As the southern gate of an expanding market in an open-door China, Hong Kong commands a strategic location for capital accumulation by the private sector. For most of the general public, earning a living is of prime importance. These economic motives are reflected in the substantial concerns about spatial development as identified in the Territorial Development Strategy Review published in 1996: restructuring the urban fabric to provide economic space for Hong Kong as a regional hub; rapid population growth and the need to provide more land for residential development; economic restructuring and the future of old industrial areas; mismatch of residential locations and employment opportunities; environment and development; improvement of cross-border links with the Chinese hinterland; and the need for more mass transit railway systems (Planning Department, 1996a).

Traditionally, the closed door and Hong Kong-centred urban planning system had served to maintain economic dynamism in the colony. With democratization introduced in the 1980s, the planning system is forced to be more open and to serve not just economic interests. However, developers have been very sensitive and opposed to the opening up of the urban planning system and the idea of entrusting more power to planners. The debate about future directions of the urban planning system is complicated by Hong Kong's reversion to Chinese rule in 1997. One can see that the urban planning system is a contested terrain. Given uncertainties about further democratization in the territory, it may be argued that urban planning will remain subservient to market mechanisms at least in the short to medium term.

3.3. THE POLITICAL ECONOMY OF SINGAPORE: DOMINANT ROLE OF THE STATE

Unlike the politically apathetic residents who treated Hong Kong as a transitional home in the early days of British rule, the Singaporeans under British colonial rule had been very active in fighting for internal self-governance, which they finally attained in 1959.* At the 1959 election, the People's Action Party (PAP) adopted a united front strategy and won all the seats in the election. After the victory, the PAP purged its left wing elements which never regained any political muscle in the PAP-dominated administrative state.

Singapore became independent from Britain in 1963 and joined the Federation of Malaysia. The merger proved to be a failure. When Singapore separated from Malaysia in 1965, Britain also announced the withdrawal of its armed forces (which

*The British, however, were still responsible for external affairs and internal security matters.

then constituted 20% of the economy and 6% of direct employment) from Singapore by 1968. The separation from the rich hinterland and the lack of local capacity to initiate industrialization, the symbol of progress and development, reinforced the 1959 elected government's two pronged development strategy: an increasingly active role of the government in the economic arena through the establishment of statutory boards, state and semi-state owned companies, and the creation of favourable climate for foreign investment. Under this strategy, the role of the state is to provide the framework within which the market mechanism can work where both the visible and invisible hand can work alongside each other in the development process (Lim, 1991, p. 203).

The government's involvement in the economic sector is facilitated by statutory boards, wholly and partially owned state companies. Government business activities, in addition to public utilities, accounted for 45% of GDP, and the state employed close to 24% of the labour force (Clad, 1989, p. 143 and FEER, 17.8.89, p. 55 quoted in Paul, 1993, p. 293). As argued by Chen (1983), statutory boards, each with its specific terms of reference and duties, play a crucial role in development planning in Singapore.* The chairman and board members are appointed by the minister, but their staff are not civil servants. In other words, they are removed from direct accountability to the legislature and are free of government procedures and centralized control (Chen, 1983, p. 7). Each board can try to achieve its objectives under the general guidelines of the minister or cabinet. 'The purpose of the statutory board structure was to emphasize efficiency and expediency in specialized economic and urban functions, without loss of ultimate responsibility to the appropriate cabinet minister' (Chen, 1983, p. 7). In fact, it has been argued that the Government of Singapore is the most important entrepreneur in the Singapore economy† (Chen, 1974, p. 17). Since 1985, the government has embarked on divesting some of its holdings in the state-owned companies (Chua, 1996, p. 95). However, the government has tried to ensure that it retains control over the company.

On the other hand, it was then believed that foreign direct investment could accelerate the industrialization process and an export-oriented economic development strategy was pursued. Similar to Hong Kong, Singapore's history as a port serviced by an established financial and service sector had facilitated its transition to an industrial centre. The government also exercised active control over the labour market to attract foreign investment. For instance, in 1968, the Employment Act and the Industrial Relations (Amendment) Act were designed to place severe limits on the workers' right to strike and give management the sole jurisdiction over recruitment (Chen, 1976, p. 37). As a result, foreign ownership share of the GDP increased from 15.7% in the 1966–1973 period to 28.1% in the 1980–1984 period (Lim and Pang,

*For example, the Housing and Development Board was established in 1960 to undertake large-scale public housing programmes; the Economic Development Board (EDB) was established in 1961 to promote industrialization and economic development; the Jurong Town Corporation was created in 1968 to provide industrial sites as well as ready-made factories to meet the increasing demand of the industrial sector.

†In 1986, there were 608 government-linked companies in Singapore which were controlled directly or indirectly by the government holding companies or statutory boards (Lim, 1991, p. 206).

1991, p. 52 quoted in Paul, 1993, p. 293); and in 1991, it reached 34.5% (Singapore Department of Statistics, 1991, p. 83).

As a single party state practising 'paternalistic dictatorship in between elections', the central role of the government is not confined to economic matters (Fig. 3). In fact, powerful political leaders have deployed 'the state machinery in a highly interventionist manner throughout the polity, economy and society' (Chua, 1996, p. 99). The state controls the institutions which articulate the interests of society, such as the labour unions, the mass media, and higher education (Paul, 1993, p. 295). The government has mobilized domestic savings through the compulsory Central Provident Fund (CPF) for public development expenditure. Employees are allowed to use their CPF to purchase their own Housing Development Board flats. In the social arena, political engagement to direct, formal process is limited and there is arguably no genuine civil society in Singapore (Rodan, 1996, p. 37). The government discourages conflict, confrontation and bargaining. While there are some non-government organizations in Singapore,* the Societies Act proscribes political activities by organizations not registered for that purpose (Rodan, 1996, p. 38). In the regulated society, people are used to following rules and regulations set by the government.

In the administrative state, the bureaucracy is responsible for the complex development activities in the export-led economy. Since mass politics is discouraged, a lot of political issues are treated as technical ones. Hence professional expertise is highly valued and the city-state depends heavily on the judgement of the leadership to plan and implement with complete and irreversible power (Chen, 1974, p. 5). Consequently, the major means of urban governance is careful planning and monitoring by the state. Put otherwise, 'norms and ends are selected by the political leadership and the bureaucracy, and they are usually supported by the elite and accepted by the masses' (Chen, 1983, p. 22). As argued by Jones and Brown (1994, p. 85), '[i]n the managerial vocabulary of Singapore politics, the elite consists of 'experts' guided by 'pragmatism', while the electorate's role is to 'endorse' and 'feed-back,' never to oppose'.

In this elite-oriented culture, efficiency, productivity, pragmatism and meritocracy are highly valued by the elite and the masses (Chen, 1983, pp. 22–23). Therefore, unlike their vocal counterparts in Hong Kong, the middle class in Singapore has a considerable stake in the PAP state and its meritocracy ideology which legitimizes their privileged social and economic positions (Rodan, 1996, pp. 20, 39). The middle class in Singapore is characterized by its *kiasu* ('afraid to lose' or 'fear of failure') behaviour, which in turn sustains the development of managerial corporatism (Jones and Brown, 1994, pp. 84, 87). The ruling philosophy of the Singaporean government is reflected in the following principles of social and political development (White Paper, 1991):

- nation before community and society above self;
- family as the basic unit of society;

*These include the Nature Society of Singapore (NSS), the Association of Women for Action and Research (AWARE) and the Association of Muslim Professionals (AMP) (Rodan, 1996, p. 37).

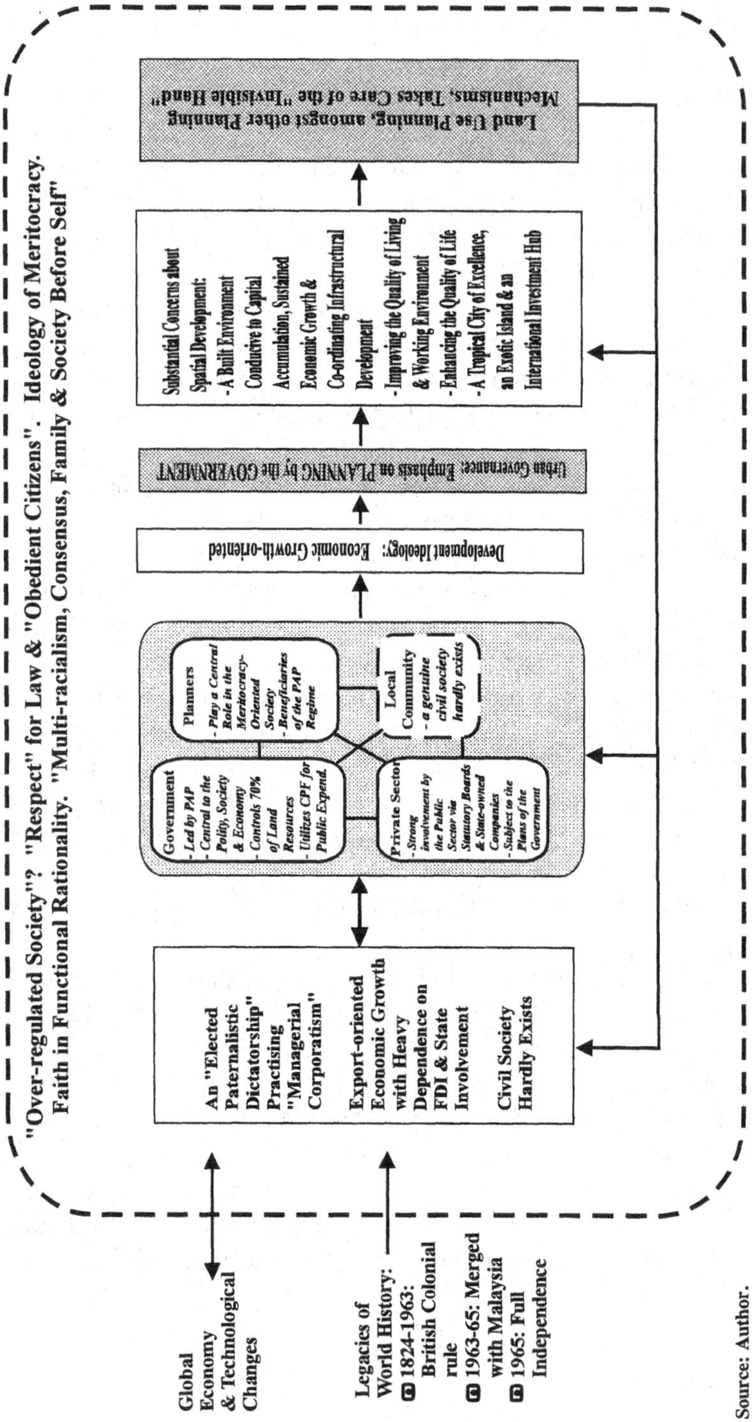

Source: Author.

FIG. 3. Political economy and urban planning: the case of Singapore. (Source: Author)

- community support and regard for the individual;
- consensus, not conflict, as the basis of national decision making; and
- racial and religious harmony.

With the increasing educational standard of the average citizen, the PAP government has probably realized that 'consensus' is essential as a basis of policy making and electoral support. For instance, the 1985 Economic Committee* consulted more than 1000 individuals in various capacities to formulate recommendations for changing the economic development strategy (Chua, 1996, p. 103). Other measures included: 'the establishment of a Feedback Unit, an extra-parliamentary institution in 1985 to take suggestions from the public and to explain government policies at the grassroots level; the adoption of Government Parliamentary Committees in 1987; the introduction of town councils; and the establishment of the Institute of Policy Studies, a 'think-tank' that attempts, among other things, to involve professionals in public policy discussion' (Rodan, 1996, p. 33).

As the owner of some 70% of land in Singapore (Urban Redevelopment Authority, 1995, p. 5) and with legislative provision which allows the government cheap acquisition of land for development, the government is able to ensure the provision of a built environment conducive to capital accumulation by foreign investors, as well as attractive to local residents. Substantial concerns about physical development can be seen in the objectives of the revised Concept Plan which aims to develop Singapore into a Tropical City of Excellence, an exotic island and an international investment hub (Urban Redevelopment Authority, 1991, p. 11):

- safeguarding and providing land for development;
- encouraging sustained economic growth and co-ordinating infrastructural development;
- improving the quality of living and working environments;
- enhancing the quality of life; and
- projecting an Asian city and tropical island image.

As central planning is the major means of urban governance, urban planning mechanisms play an important role in realizing development policies. Unlike the case of Hong Kong where market forces work together and sometimes even override urban planning, in Singapore, market forces are 'taken care' of by the visible hand of planning experts.

3.4. THE POLITICAL ECONOMY OF TAIWAN: 'GOLD-POWER' ALLIANCE VS DEMOCRATIZATION

As a colony of Japan from 1895 to 1945, Taiwan served as the agricultural appendage of the Japanese economy. Japanese investment in Taiwan's rural sector had provided a firm foundation for its subsequent industrialization when the mainlanders

*The Economic Committee was constituted during the recession years of 1985–1987. The Committee was instrumental in reorienting some of the economic policies of the government (Chua, 1996, p. 105).

arrived with their money and entrepreneurship. The taking over of Japanese properties by the Nationalist Government allowed the state to play a more important role in the initial phase of Taiwan's economic takeoff. In 1952, the public sector shared 56.6% of industrial output in Taiwan's economy (Fei *et al.*, 1979, p. 50). The government-owned enterprises then accounted for 77.7% of Taiwan's total exports. However, on the one hand, the military government from the mainland did not treat the Taiwan population as equal and on the other focused on profits and commandeering Taiwan's resources.* 'The deterioration of the political economy reached its nadir' in 1947 and resulted in an island wide revolt (Selya, 1995, p. 2). The incident resulted in the liquidation and exile of the Taiwanese nationalists and leftists. Then, in the name of anti-communism and stability, the Nationalist Party ruled Taiwan under the martial law and Temporary Provisions During Period of Communist Rebellion (Yen, 1990, p. 174). Similar to PAP in Singapore, arbitrary arrests, ruthless suppression of dissidents, harsh censorship and tight social control were employed to ensure the dominance of the one party rule.

The historical hatred developed by the native Taiwanese of the corrupt Nationalist Party, and the government it led, had exacerbated the traditional Chinese culture of paying little respect to rules and regulations which are perceived as state instruments of repression. This has severe implications on urban governance issues.

If it were not for the massive foreign aid from the United States of America (until 1965), the successful implementation of land and agrarian reforms in the early 1950s, and the pursuit of vigorous industrialization (from import substitution in the 1950s to export-oriented industrialization in the 1960s) which eventually made Taiwan an Asian NIE, the legitimacy of the mainlanders' rule over the island would be seriously in doubt. In fact, the grudges between the mainlanders and islanders did not diminish until recent years when both the Central Standing Committee of the Nationalist Party and the Central Government in Taiwan began to 'Taiwanise' and 'democratise'.

Resembling Hong Kong and Singapore, the government in Taiwan has stressed the importance of economic growth. The Nationalist Party-dominated central and local governments in Taiwan adopted a three-pronged development strategy. First, the state exerted a pro-active role in economic development. Similar to the Economic Development Board in Singapore, Taiwan also has an inter-ministerial Council for Economic Planning and Development (CEPD) to steer macroeconomic development through a series of plans of various lengths (Selya, 1995, p. 9). Major infrastructural projects were constructed under these plans.† Wade (1990) has described the Taiwanese economy as a 'governed market'. The government created its own state-run industries, such as the Chinese Petroleum Industry, by severely limiting foreign

*See Ballantine (1952), Kerr (1965).

†From 1973 to 1980, ten major construction projects such as the Sun Yat-sen national freeway, Chiang Kai-shek international airport, petrochemical industry, Taichung port, etc. were built. From 1980 onwards, there were 12 new development projects, including, for instance, around-the-island railroad, new towns and housing projects expansion, etc. From 1985 onwards, 14 key projects were launched, such as expansion of highways, rail system, telecommunications modernization, Taipei mass rapid transit system, etc. (Selya, 1995, p. 11).

companies from investing in this sector (Douglass, 1994, p. 55). In 1994, government and public enterprises comprised 46% of gross capital formation (CEPD, 1995). Second, to boost domestic industrial capacity, a '19 Point Reform' Programme was adopted in 1960 to promote exports. The Programme committed the government to encouraging saving and private investment, unification of foreign exchange rates, liberalized exchange control, tax reform, the sale of government enterprise to private owners, and holding military expenditure at the 1960 level (Hsiao, 1981, p. 251). Various industrial zones were established to meet the needs of growing small to medium sized manufacturing firms. Third, Taiwan created the first export-processing zones to attract foreign investment. Various incentives were provided: the right of foreign investors to 100% ownership, conversion and remittance of all net profits, protection against government expropriation or requisition for 20 years if the foreign investment in an enterprise is 45% or more of the total registered capital, tax holidays and tax breaks (Selya, 1995, p. 10).

As a result of the mainlander-dominated government's involvement in the economy, a close relationship developed between the rich ('gold') and the bureaucrats ('powerful'). During the authoritative years, government officials used to develop close ties with private sector business leaders through numerous state-sponsored industrial associations. Bureaucrats also used these industrial associations to conduct industrial surveys, collect and disseminate business information, solicit policy inputs, and implement government policy initiatives (Chu, 1994, p. 119). Since the business leaders were handpicked by the Nationalist Party leader, a complex web of clientelistic relationships developed which provided the party-state elite a mechanism of intimate control and selected co-optation (Chu, 1994, p. 119). This bureaucrat–business alliance has developed with growing influence in Taiwan's development process. In the Legislative Yuan of 1989–1992, 38 of the 101 popularly elected members had publicly known ties with at least one business group, with many of them holding positions such as company director and receiving free company cars, chauffeurs and secretaries (Chu, 1994, p. 126). Of the 24 magistrates and mayors elected in 1989, 18 received financial backing from one business group or another (Chu, 1994, p. 126).

The economy in Taiwan had been restructured since the 1950s. At first an agrarian economy, it shifted to export-oriented industrialization in the 1960s. Since the mid-1970s, the labour intensive industries have been upgraded and high-tech industries have been emphasized. Its manufactured products for export have shifted from low end consumer goods to high tech ones, such as computer equipment. In 1994, Taiwan was the 13th largest trading nation in the world (Mann, 1994). In recent years, Taiwan has substantially liberalized its economy: easing interest rate controls, lifting credit and exchange controls, removing trade barriers, and increasing foreign access to capital markets (Taiwan Research Institute, 1996). Ironically, as a result of economic liberalization, the growing importance of the economic elite has diminished the state's control over policy outcome. The hegemonic role of the state is not only challenged at the economic front.

Like Singapore, the Nationalist Party tried to maintain a tight grip on major voluntary groups in Taiwan. However, because of economic success, a large middle class is emerging. Unlike the middle class in Singapore who are the main beneficiaries of the meritocracy oriented policy, the middle class in Taiwan have to face the reality of a powerful 'gold-power' alliance. Like their Hong Kong counterparts, the liberal-minded middle class demands political participation and institutional reforms. In tandem with the proliferation of pressure group activities is the explosive growth of the mass media (Tien, 1989, p. 251). However, the rapid democratization process would not have been possible without political reforms initiated by the government itself since the late 1980s:*

- July 1987: Formal end of the martial law.
- January 1988: End of ban on founding of new newspapers.
- February 1988: Standing Committee of ruling Nationalist Party approved parliamentary reform and encouraged lifetime members of parliament to retire.
- January 1989: Parliament approved bill to legalize the formation of new political parties.
- April 1990: Nationalist Party approves plan for retirement of legislators elected in 1947.
- May 1991: End of Temporary Provisions During Period of Communist Rebellion which scrapped certain emergency powers that superseded the constitution.
- December 1991: Last of old deputies retired from Parliament and polls held to make body fully elected.
- June 1992: Parliament passed resolution ordering destruction of tens of thousands of dossiers on civil servants and employees of state companies. This ended checks for ideological reliability.
- January 1993: Ban on new radio stations ended.
- June 1993: Parliament ended restriction on broadcasts in Taiwanese dialect.
- March 1996: Taiwan held the first direct presidential election.

Unlike the small city-states of Hong Kong and Singapore, there is a three-tier government structure in Taiwan: the central, Taiwan provincial† and local governments. As a result of the above political changes, a decentralization of power has been witnessed. However, the ability of the local administration to cope with changing needs is limited by scarce human and financial resources. The current taxation system deprives local administration of sufficient funding (Tien, 1989, p. 132). While most of the major taxes go to the central and provincial governments, land and land-related taxes go to municipal or major city governments. County and sub-county administrative offices draw their revenues mostly from miscellaneous taxes or subsidies from the provincial government (Tien, 1989, p. 132). For instance, in 1990, 76.4% of the revenue went to the central and provincial governments (CEPD, 1991, p. 144; CEPD, 1992, p. 153). Of the 23.6% shared among city and county governments, the Taipei and Kaohsiung Municipalities absorbed almost 40% (CEPD, 1991, p. 144). This public finance system explains the inadequate provision of social amenities and facilities at the local levels. To secure finance from the Nationalist Party-dominated cen-

*Extracted from: http://www.taiwaninformation.org/politics/democ.html on 01/10/97.

†To streamline the administrative structure, Taiwan is contemplating the abolition of the provincial government.

tral government, local government officials have to rely on political loyalty and personal ties (Tien, 1989, p. 138). The system also encourages a close relationship between government officials and business interests because local governments need the financial support of the private sector. This further strengthens 'gold-power' politics (Chang, 1993a, p. 166).

Political reforms and changes have led to an interesting pattern of power alignment (Fig. 4). On the one hand, the state together with the Nationalist Party-sponsored politicians and big corporations share common political interests. On the other hand, the state together with the Nationalist Party-sponsored politicians and big corporations share common political interests (Fig. 4). On the other hand, the growing importance of the Parliament allows some progressive politicians to join force with the emerging political community to fight for various environmental, social and labour issues. One major area of concern is the quality of life issue. The long held 'growthism' ideology among the ruling elite (Selya, 1995, p. 15) and rapid industrialization in the past decades have led to serious pollution problems in the urban and rural areas. Urban areas are often 'over-developed' with few open or green spaces that are not crowded, run-down or inaccessible. These issues are politicized and have become major concerns of the growing political community.

As a result, urban governance in Taiwan is dictated by politics amongst various factions within and between the governments, the parliament, the business sector and the civil society. While there are many rules and regulations set up by different levels of government, there is no tradition of taking these laws seriously in Taiwan's society. Hence, urban governance is tainted by the 'gold-power' alliance, corrupt practice and political clientelism. The rapid democratization process since the 1980s and the increased number of elections have degenerated the 'gold-power' politics into 'black-gold-power' politics. According to the Minister of Legal Affairs, one third of the elected member in the county-level cities, a quarter of the elected members in the provincial council and 10–20% of the people's representatives in the central government have an underworld (black society) background (Chong, 1996, p. A10). In such an environment, urban planning mechanisms play a minor role in guiding the physical development of the city towards desirable ends.

3.5. CONCLUDING REMARKS

The above discussion suggests that the global–local nexus, and the socio-economic and cultural milieu have influenced the power relationship among the four major actors in the three cases under study. While the governments in Singapore and Taiwan are dominated by one party rule, Hong Kong remains largely an administrative state. Instead, economic interests heavily influence policy decision making in the territory. Although democratization has emerged since the 1980s, the impact of the political community on government decision making is still minimal. While the grip of the PAP in various socio-economic and political aspects of Singapore remains strong, the hegemonic rule by the Nationalist Party in Taiwan has been eroded by the recent democratic reforms. In other words, the government of Singapore is still

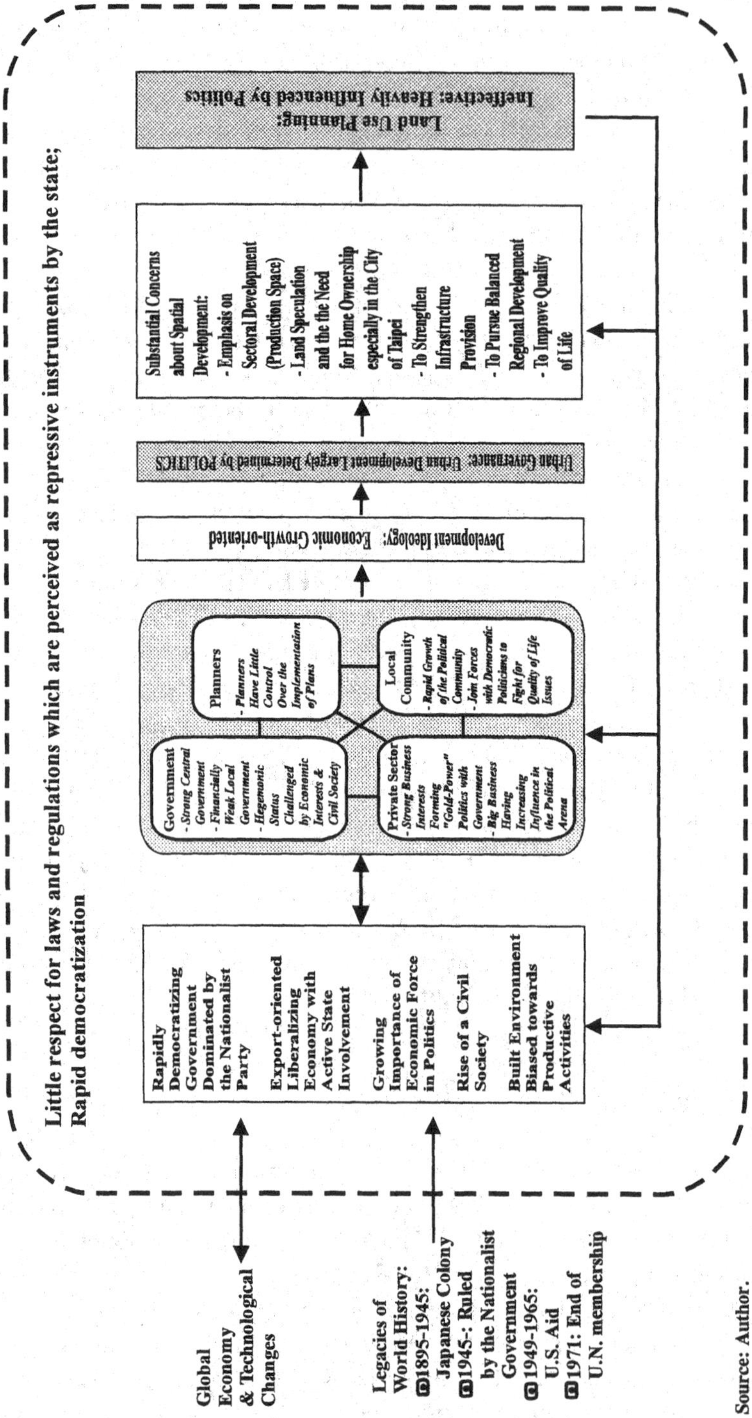

Source: Author.

FIG. 4. Political economy and urban planning: the case of Taiwan. (Source: Author)

able to direct both public and private development towards its defined goals. However, in the case of Taiwan, centralized control by the Nationalist Party has given way to a polity dominated by the Nationalist Party and challenged by opposition parties, rapidly growing economic interests and a vocal political community. The changing politics in Taiwan has reinforced the 'gold-power' alliance, especially among the Nationalist Party members and big corporate interests. How this alliance will be challenged by the opposition parties, such as the Democratic Progressive Party and the civil society, remains to be seen.

The three political economies exhibit interesting styles of urban governance. In Hong Kong, while planning is of great importance for the government (the largest landlord) in monitoring land supply and urban development, the mission of planning is basically to meet the needs of the economy. Social and environmental considerations are subordinated to the prime objective of economic growth (Ng, 1995). As the political community begins to realize the need for land use to be planned to satisfy all-round development needs, professional planners are facing an increasingly challenging task of planning to provide for both economic and life space (Friedmann, 1988).

In the case of Singapore, the PAP dominated government plays a central role in urban governance. Professional planners are the technical expert for planning rationally according to goals and objectives set by the politicians. Since the government is in charge of almost all aspects of development in Singapore, plan implementation poses few problems for planners.

Taiwan presents a more complicated story. Unlike Hong Kong, the government is dominated by the Nationalist Party. Unlike Singapore, the Nationalist Party is losing its grip on many aspects of the civil society. The dual levels of governance and the public finance system in Taiwan complicate the power networks. On the one hand, close connections between government officials and business interests are commonplace, especially at the local level as such an alliance is a major financial source for the resource hungry local governments. On the other hand, local governments also rely on their political loyalty and personal ties to higher-level governments to secure revenue. As a result, policy outcomes at different levels of governments are heavily influenced by a very dynamic network of power relationships between the government (the 'powerful') and the business interests (the 'gold'). To complicate this further is the rise of a civil society demanding an open and accountable government and a right to make decisions in urban governance. All these take place within a culture with little respect for rules and regulations. In such a fluid and rapidly changing environment, the professional planners play only a marginal role in urban governance.

The following chapters will examine in further detail the urban planning mechanisms within the three political economies.

CHAPTER 4

*Land use planning administration**

4.1. INTRODUCTION

Table 2 summarizes major land use planning administrative units in the three cases.

The land use planning administration reflects the characteristics of the respective political economies under study. In Singapore, control of land use and development is centralized and closely monitored by the Minister for National Development (who reports to the Cabinet and Parliament) with the statutory boards and government departments under its auspices. The control is also centralized in Hong Kong, but the Planning, Environment and Lands Bureau has less power than its counterpart in Singapore. The high-level Chief Secretary's Committee makes important policy decisions and the initiative to make plans and their final approval rest with the Chief Executive in Council. The Chief Executive appointed Town Planning Board and Appeal Board which, while working closely with the Planning Department, are supposed to be independent decision making bodies as regards development applications.

The case of Taiwan is the most complicated. Policy directives come from both the central and provincial governments. Different authorities within the central and local governments are responsible for different types of plans or policy areas which are not coordinated. This system is also complicated by the need for local plans to be examined and approved by upper level governments. Furthermore, the Urban Planning Commissions at all levels of governments are influential in the plan making process.

4.2. PLANNING ADMINISTRATION IN HONG KONG: FRAGMENTED CENTRALIZATION

Figure 5 shows some of the administrative units and major executive statutory boards which are responsible for, or related to, land use planning. In Hong Kong, the Chief Executive-appointed Executive Council is the highest policy decision taking body. While some of the policy proposals may be published for public consultation, the final decisions for various policies rest with the Executive Council. Before policies are sent to the Executive Council for deliberation and endorsement, they are scruti-

*It should be noted that urban planning administration systems in the three cases have changed over the years. While the author has tried to gather the most up-to-date information, readers should take note of the date when the information was obtained.

Table 2. Major land use planning administrative units in Hong Kong, Singapore and Taiwan

	Hong Kong	Singapore	Taiwan
Policy directives	Chief Executive in Council	Cabinet and Parliament	*Central Government* (Executive Yuan)
	Chief Secretary's Committee (former Land Development Policy Committee)	Ministry of National Development	Council for Economic Planning and Development-Department of Urban and Housing Development
	Planning, Environment and Lands Bureau		Ministry of Interior-Construction and Planning Administration
			Provincial Government
			Land Office
			Urban and Housing Development Bureau
Implementation	Planning Department	Urban Redevelopment Authority	*Taipei*: Department of Urban Development
	Town Planning Board	Master Plan Committee	*Kaohsiung*: Public Works Department
	Appeal Board	Development Control Committee	County and city governments and offices of county-level cities, towns and villages: Construction and Public Works Bureaux
			Urban Planning Commissions at all levels

Source: Synthesized by the author from various sources including Ministry of Interior, (1994a, p. 231.

nized by the Chief Secretary's Committee which comprises various policy secretaries. Hence, one can safely argue that the policy secretaries have the largest say in policy formulation. In some policy areas, the government has also set up advisory committees to incorporate outside opinions. However, these committees operate largely behind closed doors and policy formulation could be likened to a black-box operation in Hong Kong.

In the following, four of the bodies listed in Fig. 5 will be discussed in further detail: the high-level Chief Secretary's Committee; the Planning, Environment and Lands Bureau for making land use planning policies; the Planning Department which carries out land use planning policies; and the statutory Town Planning and Appeal Boards formed under the Town Planning Ordinance, Chapter 131 of the Laws of Hong Kong.

The Chief Secretary oversees various policy areas. One of the internal mechanisms for the Chief Secretary to supervise policy formulation is through the Chief Secretary's Committee, which is the most important committee in the government. It is chaired by the Chief Secretary and comprises all the policy secretaries to monitor government policies and related issues. One of the Chief Secretary's Sub-committees deals specifically with urban planning issues. It should be noted, however, that before 1995, the overall land development and planning policy framework was set by the Land Development Policy Committee (LDPC) chaired by the Chief Secretary.

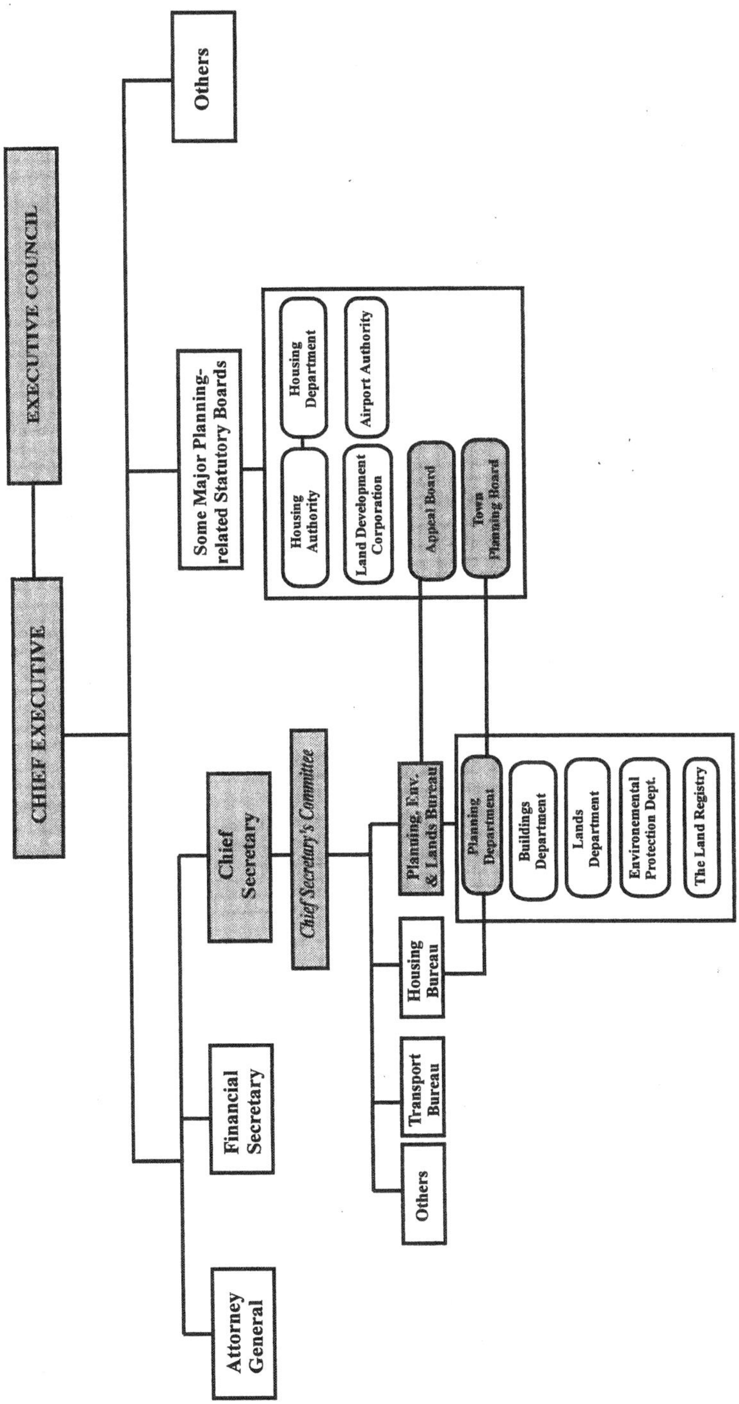

FIG. 5. Urban planning administration in Hong Kong. (Source: Author)

The policy bureau responsible for urban development and planning issues is the Planning, Environment and Lands Bureau (PELB) which provides policy directives to the Planning Department, Lands Department, Building Department, Environmental Protection Department and the Land Registry. The Planning Department was formed on 1 January 1990, through the amalgamation of fragmented planning functions previously carried out by the Town Planning Office of the then Buildings and Lands Department, the Strategic Planning Unit of PELB and various district planning sections of the then Territory Development Department. All land use planners in the government of Hong Kong are academically qualified professionals recognized by major planning institutes.* The Planning Department is responsible for formulating, monitoring and reviewing urban and rural planning policies and associated programmes for the physical development of Hong Kong. It deals with all types of planning at the territorial, sub-regional and district levels (Fig. 6). The Territorial and Sub-regional Planning Branch is responsible for medium to long-term planning strategies for the whole territory and its five sub-regions. The District Planning Branch conducts planning and development control at the local level.

As shown in Fig. 6, the Housing and Land Supply Division of the Planning Department also receives policy directives from the recently formed Housing Bureau.† The Housing and Land Supply Section within the Division is responsible for coordinating the planning for specific housing sites, the assessment of the demand and supply for public and private housing, forecasts of land requirements and supply for housing and other major land uses, and the assessment of redevelopment potential in the territory.

The Planning Department is also responsible for servicing the statutory Town Planning and Appeal Boards. Unlike the case of Singapore where statutory boards are central to translating planning intentions into reality, the functions of the Town Planning Board are limited by the Town Planning Ordinance. The Town Planning Board is a statutory body established under Section 2 of the Town Planning Ordinance. The Board consists of six official (including the Secretary of PELB and the Director of Planning) and 24 non-official members all of which are appointed by the Chief Executive. According to the Town Planning Ordinance, the Town Planning Board has two major functions: the preparation of statutory plans, and the consideration and reviews of development applications. The Appeal Board members are also appointed by the Chief Executive. Applicants aggrieved by TPB's decisions can appeal to the Appeal Board, the decision of which is final.

---

*Some land use planners are trained locally. Since 1980, the Centre of Urban Planning and Environmental Management in the University of Hong Kong has offered a Master of Science in Urban Planning programme and the course is recognized by the Royal Town Planning Institute in Britain and the Hong Kong Institute of Planners.

†The Housing Bureau was set up in November 1994 and is responsible for setting government policy on the provision of housing in the public and private sectors. It oversees the public housing programmes, monitors the operation of the private housing market, and ensures the provision of sufficient land and infrastructure to meet housing targets (Government Information Services, 1996, p. 185).

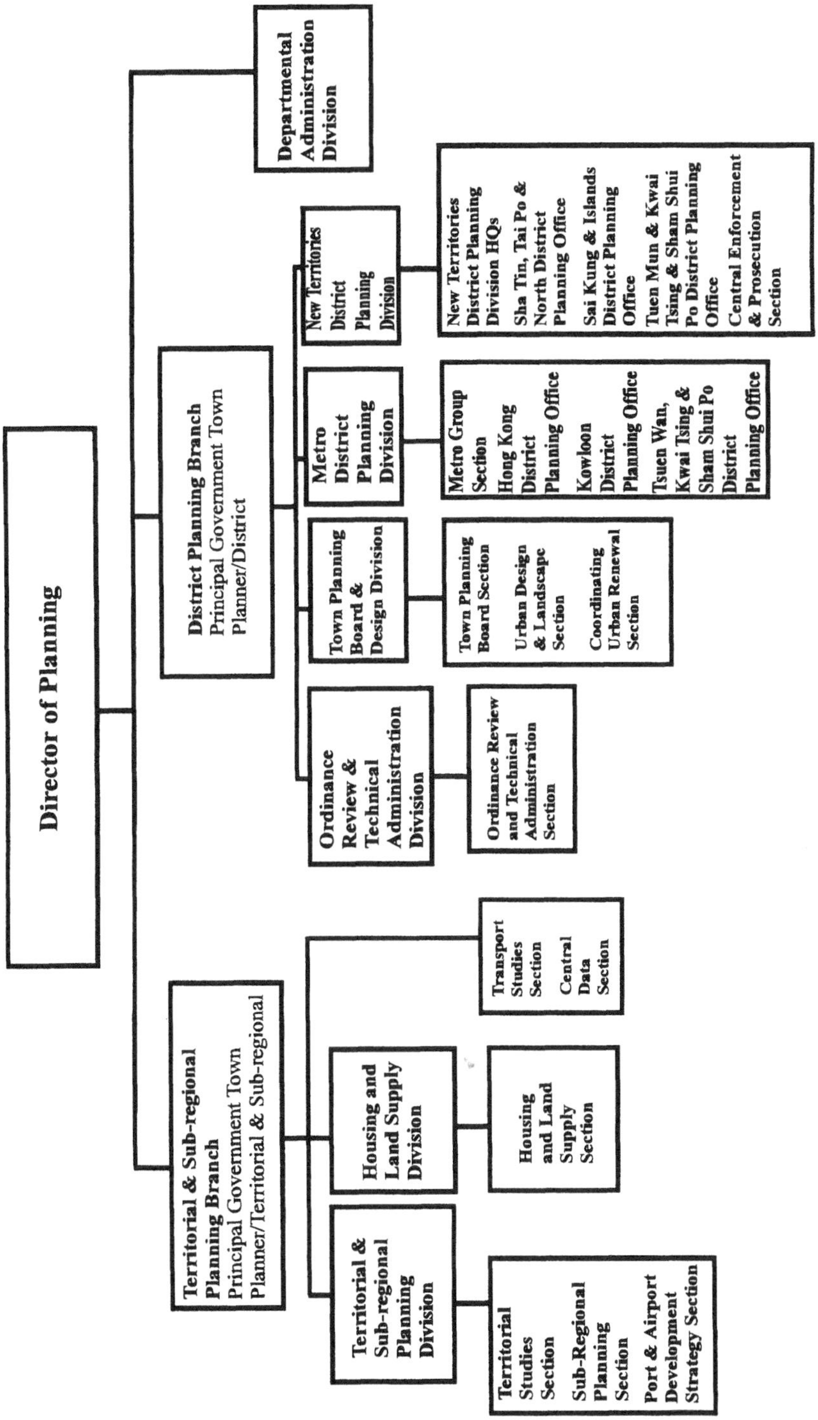

FIG. 6. Organization of the Planning Department in Hong Kong. (Source: Planning Department, 1995)

4.3. PLANNING ADMINISTRATION IN SINGAPORE: COORDINATED CENTRALIZATION

Unlike Hong Kong where land use planning-related functions are fragmented within the government, urban planning administration seems to be better organized in Singapore, as the Ministry of National Development is responsible for physical development and planning. Figure 7 shows the planning-related administrative units in Singapore. The PELB equivalent Ministry of National Development supervises four departments, five statutory boards and two professional bodies. The Ministry's major development responsibilities cover public housing, urban redevelopment, public works, and parks and recreational facilities. Its major regulatory functions include planning, development and building control. Its other functions are research and collation of statistics on the building industry and the property sector and the regulation of professional practices.* As the Ministry is in charge of the Housing Development Board, the Urban Redevelopment Authority, Public Works Department, Parks and Recreation Department, National Parks Board and Construction Industry Development Board, implementation of plans can be ensured.

The planning authority in Singapore is the Urban Redevelopment Authority (URA) which was formed on 1 September 1989 after the merging of the former Planning Department and the Research and Statistics Unit with the URA. URA is now responsible for all development control and planning functions for the physical development of Singapore, including the conservation of land and buildings (Khublall and Yuen, 1991, p. 32). Unlike land use planners in Hong Kong, the planners are not required to have professional planning degrees. URA's main functions include three aspects: planning, facilitating and regulating. URA has a crucial role in preparing and review of the statutory Master Plan, the long-range Concept Plan and the more detailed local area plans for physical development. It also facilitates development through the sale of sites programme, public sector development coordination and project services. Through various development control mechanisms, URA regulates private development in the small territory. Figure 8 shows the organization chart of URA.

According to Khublall and Yuen (1991, p. 35), the Chief Planner in URA is required to act in consultation with two advisory committees appointed by the Minister of National Development: the Master Plan Committee (MPC) and the Development Control Committee (DCC). The MPC is chaired by the Chief Planner and comprises members representing major public development authorities such as the URA, the Housing and Development Board, the Economic Development Board, the Jurong Town Operation, Land Office, Public Works Department, Ministry of Defence, Ministry of Environment, etc. to consider and recommend development projects of government departments and statutory boards. The Development Control Committee comprises members representing professional bodies of the Singapore Institute of Planners and the Singapore Institute of Architects and relevant government depart-

*The Ministry of National Development is also responsible for primary production, the import and export of agricultural and animal produce, etc.

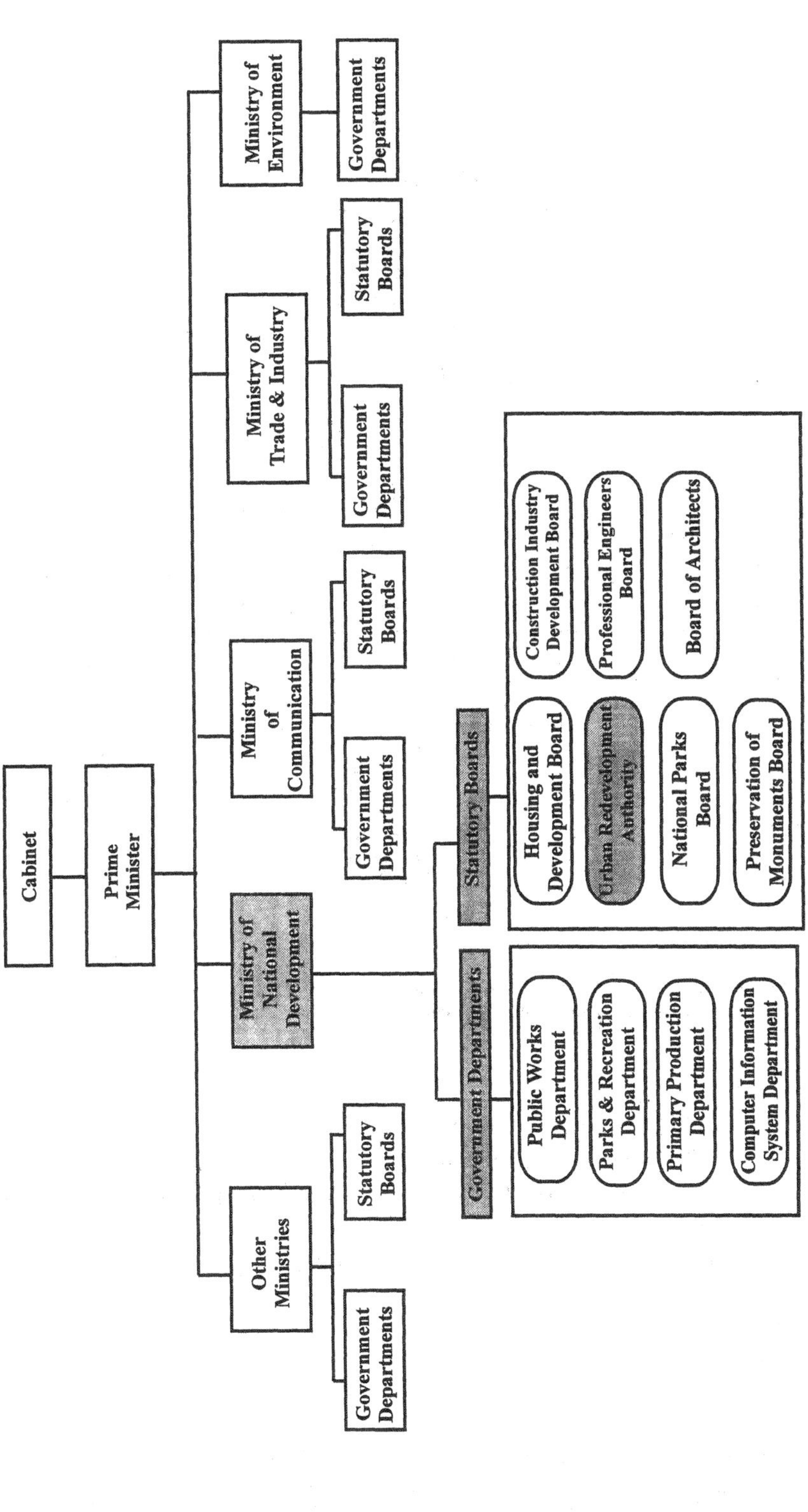

FIG. 7. Urban Planning Administration in Singapore. (Source: Based on text from the Singapore Government Internet Web Site: http://www.gov.sg in January 1997)

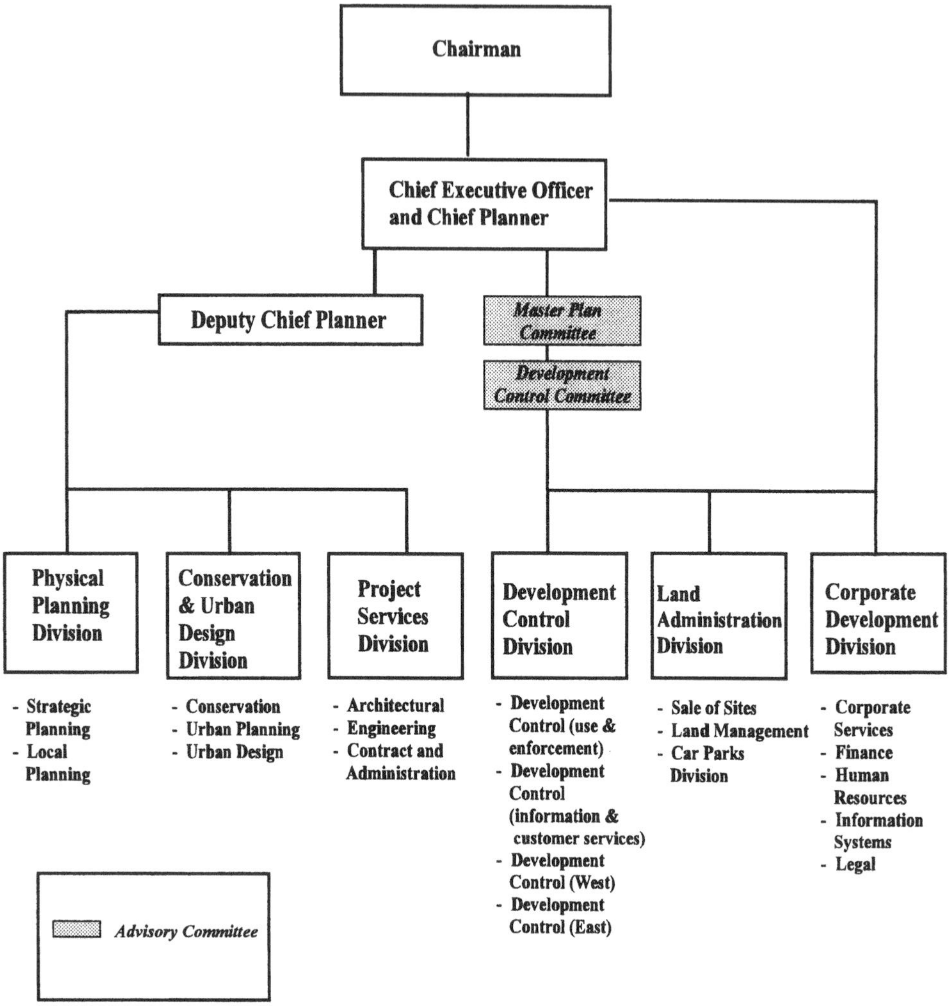

FIG. 8. Organizational chart of the Urban Redevelopment Authority in Singapore. (Source: Modified from Urban Redevelopment Authority, 1995, pp. 8–9)

ments. Although the DCC has power to approve certain categories of development, its major function is to consider and make recommendations on major development applications submitted by the private sector (Khublall and Yuen, 1991, p. 36). It also formulates, and reviews periodically, planning standards, policies and requirements related to private development proposals (Khublall and Yuen, 1991, p. 36).

4.4. PLANNING ADMINISTRATION IN TAIWAN: UNCOORDINATED FRAGMENTATION

Different from the one-tier government structure in Hong Kong and Singapore, the urban planning institution in Taiwan is complicated by the three-tier administrative structure: the central government, the provincial government and centrally-administered municipalities of Taipei and Kaohsiung, and the county (18) and city (5) level governments. Figure 9 shows the units which are related to land-use planning at the central, provincial and municipal levels. Unlike Singapore, where the Ministry of National Development is responsible for physical development and planning, national, regional, urban and various sectoral planning duties in Taiwan are fragmented among various departments within and between different levels of governments.

At the central level, the Executive Yuan under the President supervises, among others, the Council for Economic Planning and Development (CEPD) and the Ministry of Interior. Under the CEPD, the Department of Urban and Housing Development is responsible for preparing the comprehensive development plan for Taiwan and the study, examination, coordination and promotion of urban and regional development, housing construction, environmental protection and related issues. The Construction and Planning Administration (CPA) under the Ministry of Interior is responsible for regional planning, urban planning, national park planning, building regulations and public works. It also services various commissions such as the Regional Planning Commission, Urban Planning Commission, National Park Commission, Central Public Housing Fund Commission, Coordination Commission for Promotion of Land Acquisition for Urban Public Facilities, etc. The CPA is currently revising Taiwan's four regional plans; assisting county and city governments to complete local master plans; preparing an integrated plan for Taiwan's coastal areas; and updating regulations and policies such as those governing development applications in non-urban land, etc.

At the provincial level, the Construction Bureau, the Land Office, and the Housing and Urban Development Bureau, the Urban and Regional Planning Commission are related to urban planning. The fourth branch within the Construction Bureau is responsible for urban and regional planning, development and construction management (Executive Yuan, 1991, p. 275). The Land Office is responsible for land use planning, regional planning, planning for non-urban areas, equalizing land rights, setting land price and re-planning or re-adjustment of rural and urban land.* The Urban and Housing Development Bureau administers public housing construction, urban development, new towns development, infrastructure development, regional planning, special district planning and mass transit systems at the provincial level. It can be seen that there are considerable overlaps between the functions of urban planning related administrative units at the provincial and central levels. In fact, the Taiwanese National Assembly is currently planning to freeze the activities at the provincial level and suspend its next elections.

For the centrally administered municipalities, the Department of Urban Development (Taipei) or the Urban Planning Office under the Department of Public Works (Kaohsiung) and the Urban Planning Commissions are responsible for urban planning issues. In the city of Taipei, the Department of Urban Development is re-

---

*Land issues are important factors affecting plan implementation in Taiwan and will be elaborated on in later chapters.

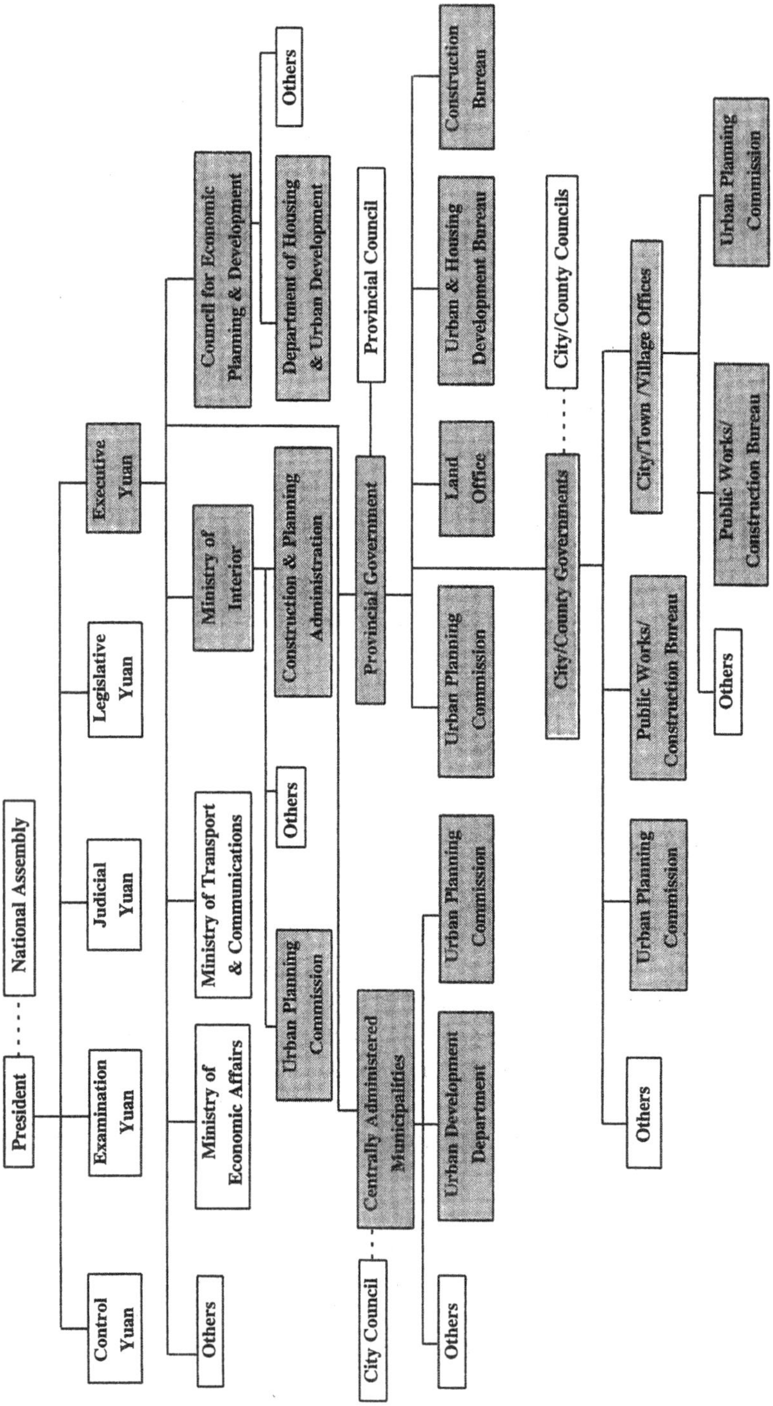

FIG. 9. Urban Planning Administration in Taiwan. (Source: Author)

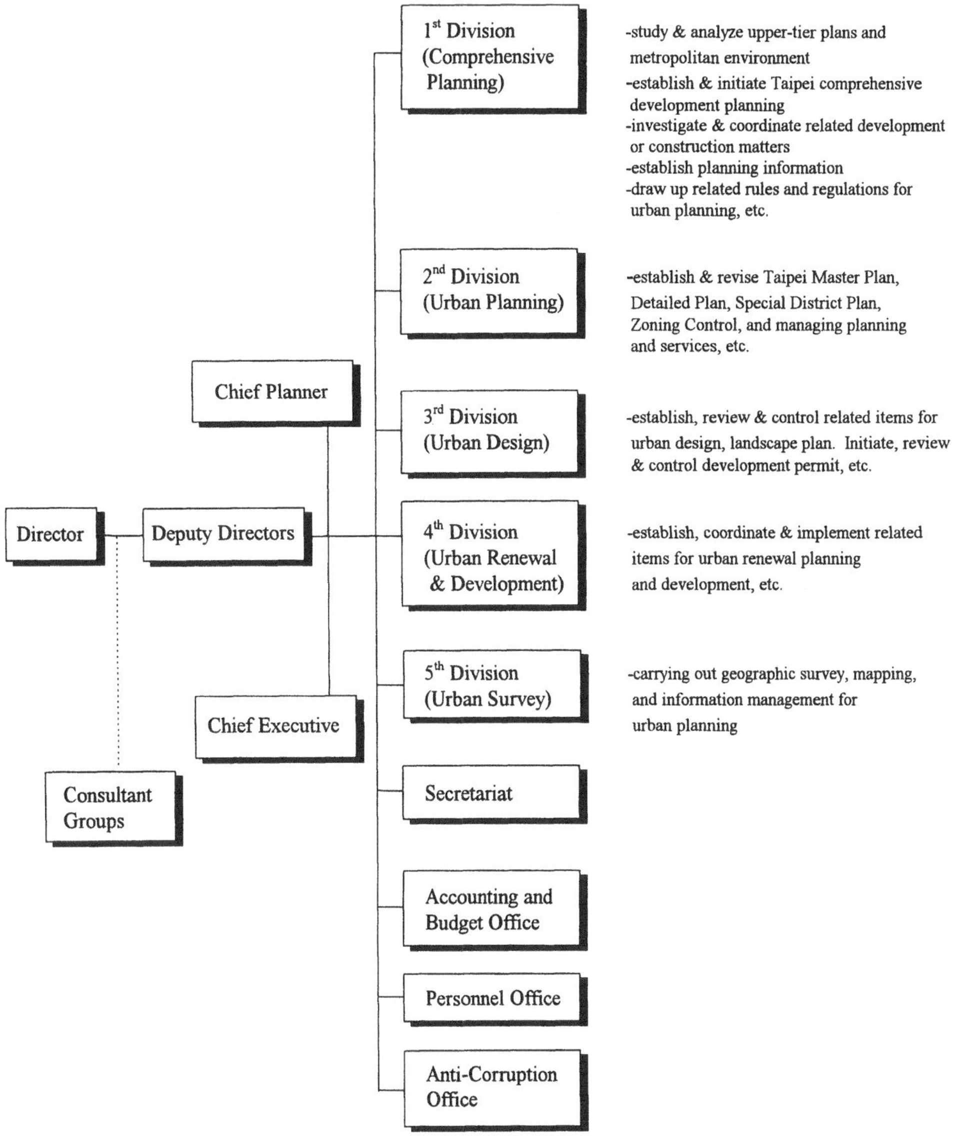

FIG. 10. Organization of the Department of Urban Development, Taipei City Government. (Source: Department of Urban Development, Taipei Municipal Government, 1994b)

sponsible for preparing the master plan, detailed and other plans; urban design; planning for urban renewal and development; and collecting data and carrying out survey for planning purposes (Fig. 10). Unlike the Town Planning Board which is serviced by the Planning Department in Hong Kong or the Development Control Committee within the Urban Redevelopment Authority in Singapore, the Urban Planning

Commissioners are appointed by the mayors and work directly under the city government.

Similar establishments are found in the city and county level governments and offices of county-level cities, towns and villages. However, as argued above, local level governments lack human and financial resources for preparing and implementing plans. Unlike the central, provincial or centrally administered municipalities' government officials, many of whom have planning degrees, it is doubtful if local governments can afford the service of well-qualified professional planners.

CHAPTER 5
Hierarchy of plans

5.1. INTRODUCTION

Table 3 highlights the hierarchy of land use plans in the three economies.

It can be seen that not all of the plans are statutory in nature, nor do they require citizen input. As will be discussed, the strategic and local plans in Singapore are closely linked. In Hong Kong, the links among the three tier plans are not explicit, as some of the strategies and plans are prepared administratively and not published. The system is the most complicated in Taiwan, where loosely related plans are made for different geographical levels.

5.2. PLANS IN HONG KONG: A THREE-TIER HIERARCHY

Hong Kong has a three-tiered system of land use plans (Fig. 11). The highest level of the hierarchy is the Territorial Development Strategy which maps out the overall land development strategy to meet the long term socio-economic needs of Hong Kong. The second tier is the sub-regional development strategies which translate territorial development goals into more specific planning objectives for the five subregions in Hong Kong. At the district level, there are four types of plans and only two of them are statutory. In other words, only two types of district plans in the whole planning hierarchy are statutory and involve mandatory citizen participation in their preparation. The rest of the plans are prepared administratively.

The outline zoning plans (OZPs) are one of the statutory plans prepared under the Town Planning Ordinance. The OZPs show the proposed land use and major road systems of individual planning scheme areas. OZPs serve to reflect existing land uses, planning intentions of future development, and special development needs. Attached to the plan is a Schedule of Notes setting out the uses which are always permitted and those which require Town Planning Board's permission (Planning Department, 1995, p. 12). The Development Permission Area (DPA) Plans are products of the 1991 amendment of the Town Planning Ordinance to extend its control over the whole of the New Territories. DPA plans are interim plans prepared for areas not yet covered by outline zoning plans. Both OZPs and DPAs can be bought from the government. The larger-scale plans such as the outline development plans and the layout plans serve better purposes in terms of development control, but they are regarded as departmental plans used mainly for administrative purposes within the government.

Table 3. Hierarchy of plans in Hong Kong, Singapore and Taiwan

	Hong Kong		Singapore		Taiwan	
	Name of plans	Statutory?	Name of plans	Statutory?	Name of plans	Statutory?
National	N.A.	N.A.	N.A.	N.A.	Comprehensive Development Plan for Taiwan	N
Regional	N.A.	N.A.	N.A.	N.A.	Regional Plans: East, South, West and North	Y
County region	N.A.	N.A.	N.A.	N.A.	Metropolitan or Joint Development Plan	N
					County and City Joint Development Plan	
City-wide urban land use plans						
Strategic	Territorial Development Strategy	N	Concept Plan	N	Master Plans	Y
			Master Plan	Y		
Sub-regional	Sub-regional Development Strategies	N	N.A.	N.A.	N.A.	N.A.
District	Outline Zoning Plans	Y	Development Guide Plans (finalized as Master Plan Proposals)	Y	Detailed plans	Y
	Outline Development Plans	N	Urban Design Structure Plans	N		
	Layout Plans	N	Conservation Master Plan	N		

Source: Author.

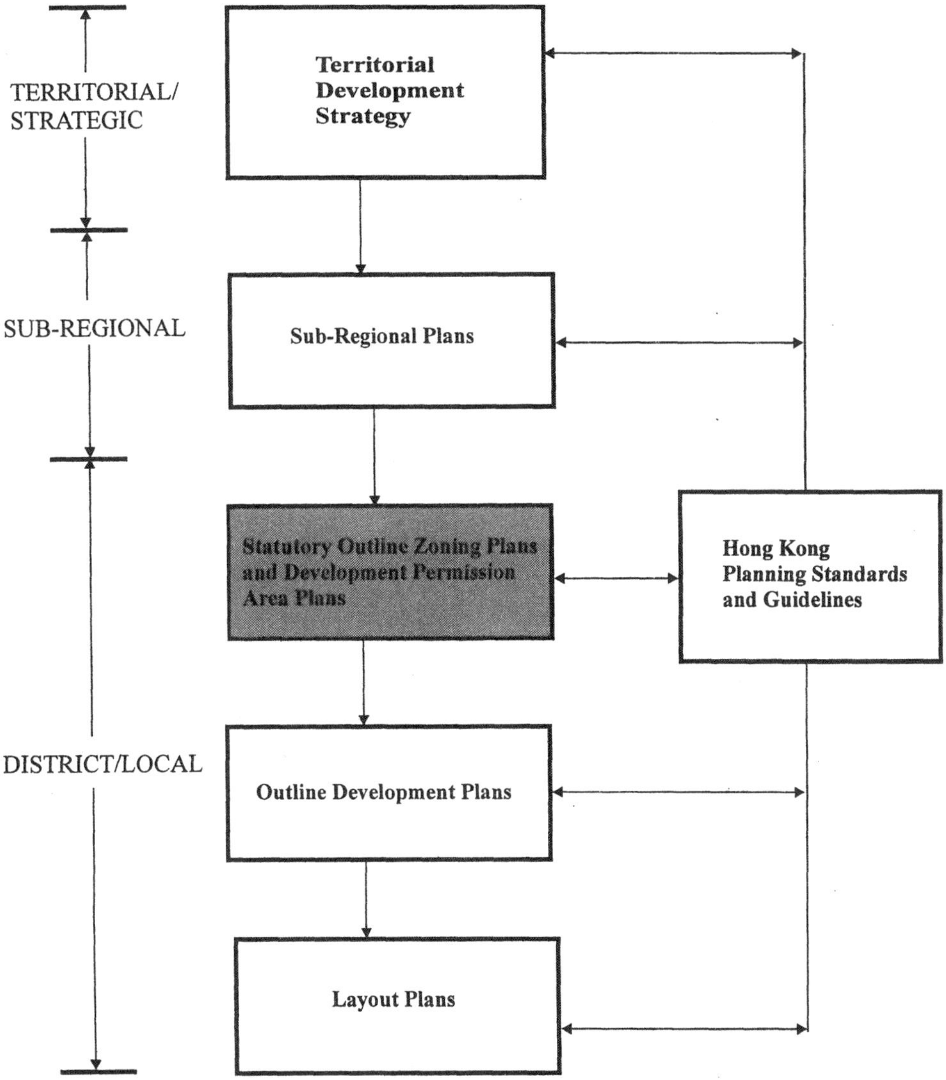

FIG. 11. Hierarchy of plans in Hong Kong. (Source: Planning Department, 1995)

The Hong Kong Planning Standards and Guidelines serve as a policy document to guide the preparation of the above plans. The document sets out provision standards, criteria for site reservation, locational factors and site requirements of various land uses. It also provides guidelines for the preparation of planning briefs for development projects in both public and private sectors (Planning Department, 1995, p. 16). This important document, which affects plan making at all levels, is prepared and reviewed by the administration and there is no public input to it.

Cuthbert (1991, p. 583) has argued that by leaving out the most important strategic plans from the statutory framework, the 'government has obviated the responsibility of being held accountable for an overall development plan. Bureaucratic discretion has been optimized along with the capacity to accommodate shifting patterns of investment and economic interests'. Even when the government had tried to consult the general public in the formulation of some of its sub-regional development strategies such as the Metroplan (1988–1991) and the review of the Territorial Development Strategy in 1993 and 1996, the extent that public representation affected the final outcome cannot be ascertained. Given the fact that the average citizen does not have much power to influence broader policy issues which constitute the basic parameters for strategic planning, it is unlikely that the general public would question fundamental assumptions and hence the consequent recommendations.

5.3. PLANS IN SINGAPORE: A TWO-TIER HIERARCHY

Singapore's development is guided by two instruments: the long-range Concept Plan and the statutory Master Plan within the purview of the Urban Redevelopment Authority under the Ministry of National Development (Foo, 1992, pp. 104–105). The statutory Master Plan provides the legal framework for control over private sector development, while the non-statutory Concept Plan guides public sector development (Khublall and Yuen, 1991, p. 39).

The Master Plan was first formulated during 1952–1955, and approved by the Government in August 1958. The Master Plan has since undergone six reviews and various alterations and the last review was carried out in 1990 (Urban Redevelopment Authority, 1993, p. i). The Master Plan originally aimed to regulate land uses through zoning, density/plot ratio control and reservation of land for infrastructure development (Khublall and Yuen, 1991, p. 40). However, because of dynamic development in Singapore, the Master Plan only depicts what has happened in the past and does not indicate future development possibility (Khublall and Yuen, 1991, p. 42). As a result, a new approach of planning has been adopted since 1988 (Urban Redevelopment Authority, 1993, p. i).

In the new planning system (Fig. 12), the Concept Plan maps out the long term land use and development strategy for the year 2000 and beyond, whilst the Development Guide Plans (DGPs) translate the intentions of the Concept Plan to guide development at the local level (Urban Redevelopment Authority, 1993, p. i). The Concept Plan is equivalent to the Territorial Development Strategy in Hong Kong and represents a long-term physical development strategy for Singapore. The Concept Plan was revised in 1991 to accommodate an eventual population of four million and embodied the following goals:

- Satisfy all domestic functional needs with improved quality of environment and quality of life.
- Stay competitive with other leading cities in economic growth.

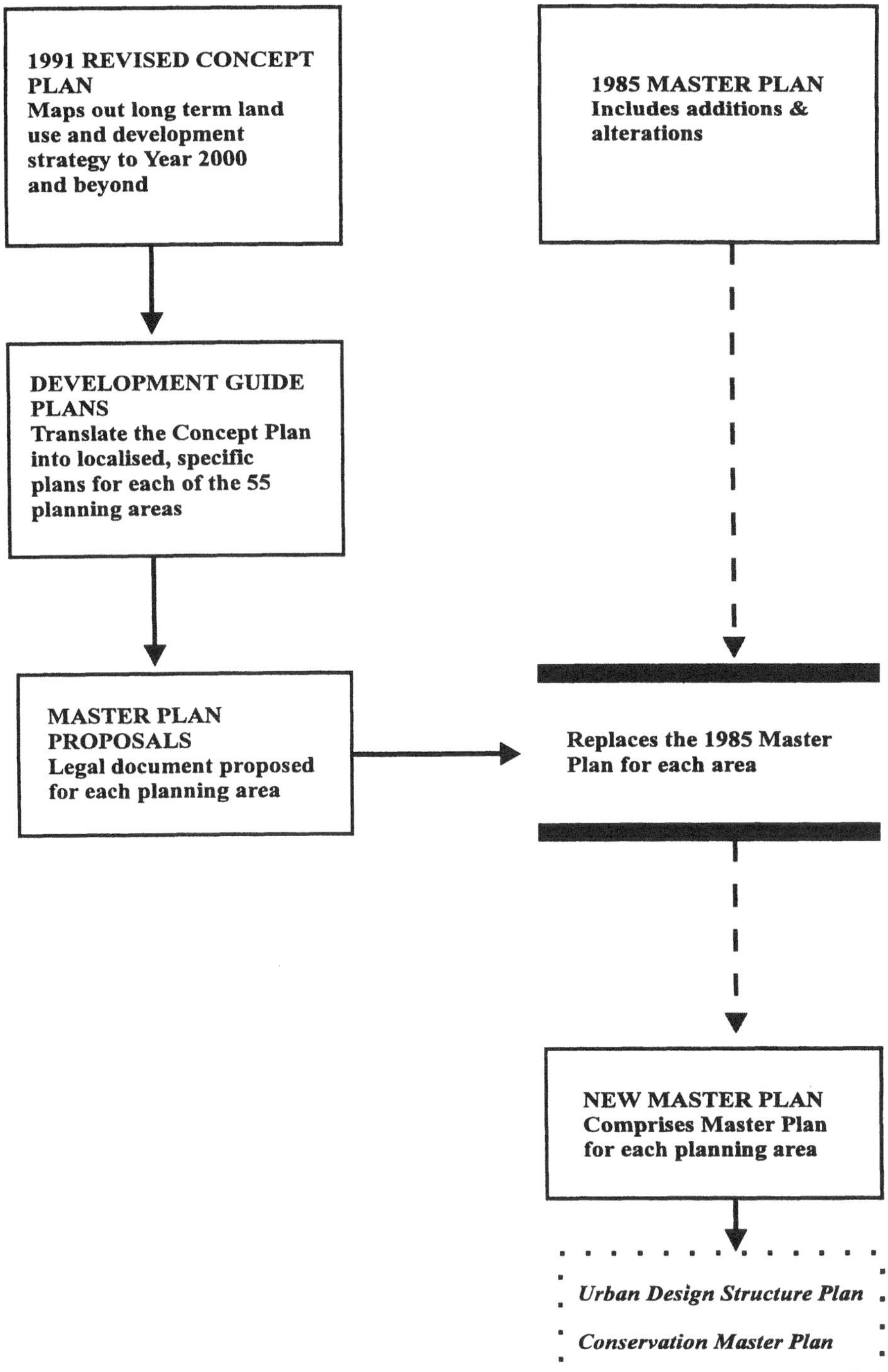

FIG. 12. Hierarchy of plans in Singapore. (Source: Urban Redevelopment Authority, 1994, 1996)

- Ensure efficient mobility of people and goods to support urban and economic development.
- Create an enduring beautiful and unique Asian city, taking full advantage of geographic, historic, climatic and cultural assets (Koon, 1995, p. 116).

Under the Concept Plan there is a second layer of more detailed Development Guide Plans (DGPs) for each of the 55 planning areas in Singapore. The DGPs provide the planning vision for an area and control parameters to guide development such as land use, intensity and height (Koon, 1995, p. 119). Unlike previous plans made by the URA, DGPs are exhibited for public comments and feedback is also sought at dialogue sessions. Some DGPs are even prepared by the private sector. These provide new and constructive ideas for URA to finalize and approve the DGPs. DGPs also serve an important function of linking the Concept Plan and the statutory Master Plan. Each approved DGP becomes the Master Plan for the specific area concerned. When all the 55 DGPs are completed, they form the new overall Master Plan guiding Singapore's development in more detailed terms. The system is similar to the case of Hong Kong. While the Territorial Development Strategy (TDS) is not statutory in nature, the district level OZPs are made within its policy framework. While the public are not involved in the formulation of the Concept Plan, they do have a chance to express their opinions on the revised Master Plans when the DGPs are incorporated into Master Plan proposals.

Other non-statutory plans in Singapore include urban design master plans at district and project levels. Urban design structure plans deal with matters such as building edges, pedestrian linkages, open spaces and the use of different building height zones which are all incorporated in the URA sales of sites programme (Urban Redevelopment Authority, 1996). Furthermore, the conservation master plan provides a systematic framework to identify buildings and areas for conservation and preservation.

5.4. PLANS IN TAIWAN: A FOUR-TIER HIERARCHY

There are four tiers of loosely linked plans in Taiwan (Fig. 13). The non-statutory national comprehensive development plan for Taiwan is prepared by the Department of Urban and Housing Development under the Council for Economic Planning and Development. The comprehensive development plan aims to promote a reasonable distribution of population and economic activities; to improve the living environment; and to conserve and develop natural resources (CEPD, 1987, p. 2).

There are four regional plans in Taiwan. The making of regional plans is stipulated in the Regional Planning Law and its Implementation Details. The East and Central regional plans are prepared by the Urban and Housing Development Bureau at the Provincial Level and the North and South regional plans are made by the Construction and Planning Administration under the Ministry of Interior at the central level. Similar to the national comprehensive development plan for Taiwan, regional plans are for promoting the utilization of land and natural resources; a

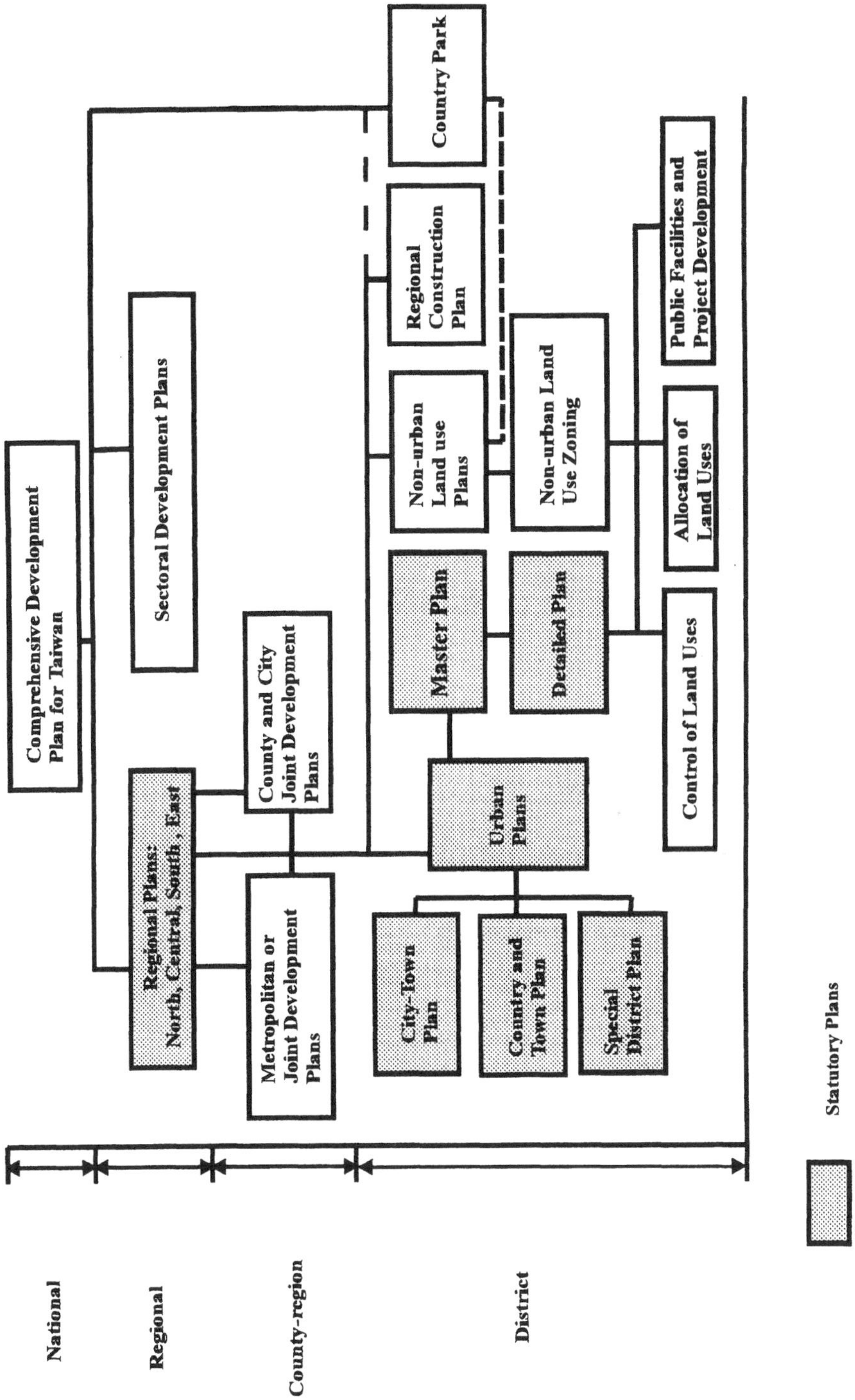

FIG. 13. Hierarchy of plans in Taiwan. (Source: Modified from Xin, 1991)

reasonable distribution of population and production; acceleration of an all-round economy; improving the living environment and social welfare (Department of Urban Development, Taipei Municipal Government, 1994b, p. 335).

The county and city joint development plans are not statutory in nature. These plans are made to promote economic development, improve the living environment, and allocate land for future development of public facilities, transport and communication, education, medical services, tourism, production and social welfare, etc. (Ministry of Interior, 1994a, p. 29).

At the local levels statutory plans are made under the Urban Planning Law. There are three types of urban plans: city (town) plans, country and town plans and special district plans. Within each of these urban plans, two tiers of plans have to be prepared according to the Urban Planning Law: the master plan and the detailed plan. Except for the difference in scale, both the master and detailed plans are zoning plans. In other words, unlike the Territorial Development Strategy in Hong Kong or the Concept Plan in Singapore, Taiwanese cities lack strategic land use development strategy. Although in theory the land use planners in cities should receive policy directives from higher level governments, the need to integrate local plans with higher level plans is not specified in the Urban Planning Law. As a result, the linkage between different levels of plans is rather tenuous, if not absent.

CHAPTER 6

Land use planning legislation and the plan making process at the municipal level

6.1. AN OVERVIEW

Table 4 below lists the major legislation related to planning making, development control and implementation in the three economies.

6.2. URBAN LAND USE PLANNING LEGISLATION

Table 5 lists the major sections within the land use planning legislation in the three economies. It can be seen that while Singapore's Planning Act focuses on development control provision, the Urban Planning Law* in Taiwan is mainly concerned with plan making and plan implementation by the government. The Town Planning Ordinance in Hong Kong stipulates both plan making and development control procedures. It is interesting to note that in the case of Singapore, there is no definition of planning. However, the meaning of 'develop' is carefully defined in Section 12 of the Planning Act.

6.2.1. Town planning ordinance in Hong Kong: procedures for plan making and development applications

As mentioned previously, only two of the four district plans in the three-tier planning hierarchy in Hong Kong are statutory in nature. The making of these two types of district plans is provided in the Town Planning Ordinance (TPO). The TPO is a piece of procedural legislation (Bristow, 1984). Section 2 of the TPO provides for the appointment of the Town Planning Board consisting of official and non-official members appointed by the Chief Executive. The plan making functions of the Board are specified in Sections 3 and 4. Sections 5 to 13 are related to the plan making processes and will be further discussed in the following section. Section 14 of the TPO provides that the Chief Executive in Council may make regulations for the purpose of facilitating the work of the Board. The Town Planning Board is also responsible for preparing development permission area plans as specified in Section 20.

*No official English version of the Urban Planning Law is available at the moment of writing. The English translation was done by the author.

Table 4. Major legislation related to land use planning, development control and implementation in Hong Kong, Singapore and Taiwan

	Hong Kong	Singapore	Taiwan
Plan making			
National	N.A.	N.A.	None
Regional	N.A.	N.A.	Regional Planning Law Implementation Details of Regional Planning Law
Municipal urban areas	The Town Planning Ordinance	The Planning Act (including, among others, the Draft Master Plan Written Statement 1993)	The Urban Planning Law
Development control			
Urban areas	The Town Planning Ordinance Buildings Ordinance Building (Planning) Regulations Environmental legislation Hong Kong Airport (Control of Obstructions) Ordinance	The Planning Act Draft Master Plan Written Statement 1993 The Planning (Master Plan) Rules 1991 The Planning (Use Classes) Rules 1981 The Planning (Development) (Amendment) Rules 1983	The Urban Planning Law Measures to Handle Illegal Construction and related laws Management Rules for the Construction Industry
Non-urban areas	The Town Planning Ordinance Buildings Ordinance Building (Planning) Regulations Environmental legislation	The Planning (Development) Rules 1981 The Planning (Development Charges) Rules 1989 The Planning (Development of Land Authorization) Notification 1963 The Planning (Exemption) (Revocation) Notification 1984	Rules Governing Non-Urban Land Uses Regulations on the Conservation and Use of Slopes
Implementation			
Land acquisition	The Town Planning Ordinance Crown Lands Resumption Ordinance	The Planning Act Land Acquisition Act	The Urban Planning Law The Land Law The Statue for the Enforcement of the Equalization of Urban Land Rights

Source: Synthesized from various sources by the author.

Table 5. Land use planning legislation in Hong Kong, Singapore and Taiwan

	Hong Kong: Town Planning Ordinance (Chapter 131)	Singapore: Planning Act (Chapter 232)	Taiwan: Urban Planning Law
Enactment history	Originally 1939 (Cap. 131, 1950)-26 of 1956, 3 of 1958, 59 of 1969, 59 of 1974, 37 of 1982, R. Ed. 1988, 4 of 1991, 101 of 1991, 8 of 1993	Originally Planning Ordinance 12 of 1959. Planning (Amendment) Ordinance 1964. 1970 Ed. Cap. 279. Revised Edition 1990	Originally 1939; revised in 1964, 1973 and 1988
Major sections	26 sections	41 sections	87 sections
General: short titles, interpretations, etc.	Section 1	Part I: Sections 1–5	Chapter 1: Sections 1–8
Plan making and review	Sections 2–15, Section 20	Part II: Sections 6–9	Chapter 2: Sections 9–31
Development control	Sections 16–19, Sections 21–26	Part III and IV: Sections 10–41	Chapter 8: Sections 79–80
Plan implementation by the government	N.A.	N.A.	Chapter 3: Sections 32–41. Chapters 4–7: Sections 42–78. Chapter 9: Sections 81–87
Purpose of the legislation	To promote the health, safety, convenience and general welfare of the community by making provision for the systematic preparation of approval of plans for the lay-out of areas of Hong Kong as well as for the types of building suitable for erection therein and for the preparation and approval of plans for areas within which permission is required for development.	An Act to provide for the planning and improvement of Singapore.	Section 1: This law is formulated to improve the living environment of the residents and to promote balanced development in the planning of cities, towns and villages. Section 3: In this Law, 'urban planning' means the planned development, within specific areas, of the infrastructure important for urban economic activities, transport, health, security, national defence, education and research, etc., and the rational planning of land uses.

Source: Hong Kong (1994). Singapore, (1990). Department of Urban Development, Taipei Municipal Government (1994a, pp. 1–20).

Another function of the Town Planing Board is to process development applications (Sections 16 and 17). Hong Kong adopts a partially regulatory and partially discretionary development control system. For land use development falling within column two of the Schedule of Notes accompanying the statutory layout plans, development applications have to be made to the Town Planning Board. If the applicant is aggrieved by the decisions made by the Town Planning Board, the applicant can appeal to the Appeal Board, the constitution of which is stipulated in Section 17A. The existing TPO bears no enforcement power in the urban areas. However, Sections 21–24 provide enforcement power to the planning Authority (Director of Planning) in areas covered by the development permission area plans. As discussed earlier, this is the legacy of Hong Kong's planning history. Before 1991, land use planning legislation only applies to the main urban areas. However, with the economic integration between Hong Kong and China as a result of the Open Policy in China and economic restructuring in Hong Kong, the rural New Territories has become a 'dumping' ground and 'back-up' areas of the rapidly expanding container terminals which service the export of Hong Kong financed manufacturing products made in China. To control the spread of the 'container country' in the rural New Territories, the government amended the Ordinance in 1991 so that the control of the TPO can extend to the rural areas. In order to have effective control, enforcement provisions were included in the amended Ordinance.

Immediately after the amendment in 1991, the government of Hong Kong published a comprehensive review of the TPO which resulted in the Town Planning White Bill in 1996. The Bill was published in 1996 for public consultation. The new Bill consists of six parts with 81 sections:

Part I: Preliminary (Sections 1–2)
Part II: Establishment of Town Planning Board (Sections 3–6)
Part III: Preparation of Draft Plans (Sections 7–32)
Part IV: Planning Control (Sections 33–46)
Part V: Building Works and Planning Certificates (Sections 47–52)
Part VI: Unauthorized Development (Sections 53–56)
Part VII: Appeal Board (Sections 57–68)
Part VIII: Miscellaneous

The new Bill aims at providing more channels for citizen participation and at making the planning and development application mechanisms more transparent. The amendment of the Town Planning Ordinance reflects the changing political climate and the trend towards open government in the territory. As can be expected, the development industry is strongly against the proposed changes because it believes that the introduction of more citizen participation and the increase of 'power' in the hands of government officials will bring more uncertainties (and hence more costs) to its development projects. As Hong Kong is now a Special Administrative Region of China, it remains to be seen whether the controversial Bill will pass into law.

6.2.2. Planning Act in Singapore: emphasis on development control

Khublall and Yuen (1991, p. 266) commented that '[m]ost of the provisions [of the Planning Act] relate to the recovery of money rather than the substantive matter of planning'. In fact, only four out of 41 sections of the Planning Act concern plan preparation. Part I (Sections 1–5) of the Planning Act defines some of the terms used and the appointment of the competent authority by the Minister for National Development. Part II (Sections 6–9) provides for the review of, alteration and addition to the Master Plan; and the preparation of more detailed plans by the competent authority. Part III (Sections 10–31) concerns the development of land and conservation. Singapore has adopted a discretionary development control system since 1 February 1960. Part III provides that all development and subdivision of land require the written permission of the competent authority. It stipulates the rights and obligations of the landowners or developers and the competent authority; enforcement of planning control and conservation mechanisms (Khublall and Yuen (1991, p. 21). Part IV (Sections 32–41) concerns development charge for written permission which permits 'extra' development when compared with parameters specified in the Master Plan.

Unlike the Town Planning Ordinance in Hong Kong, there is no provision for an independent board to vet development applications nor an appeal board to handle aggrieved cases in development applications. The competent authority or the Minister for National Development is responsible for processing and deciding on development applications, and on appeals against planning decisions or development charge orders. As we shall see, the Planning Act enables the Minister to have a strong control over development in the Island state.

6.2.3. The Urban Planning Law in Taiwan: emphasis on plan making and implementation

While the Planning Act in Singapore is detailed in stipulating development control mechanisms and brief in plan making processes, the Urban Planning Law in Taiwan is detailed in plan making processes and short in specifying development control. Taiwan does not have a discretionary development control system in the urban areas. Different from Singapore, Taiwan adopts a regulatory approach to control development and therefore the plans, the blue-prints, are seen as the most important tool to guide future development. As a result, there is no provision for development applications. Instead, procedures to amend the blue-prints are specified.

The whole of Chapter 2 (Sections 9–41) provides for the preparation, amendments, publication and implementation of urban plans which include the master and detailed plans. Except for the two sections in Chapter 8 concerning offences (illegal construction and enforcement provision), the rest of the Urban Planning Law stipulates detailed provisions for: land for public facilities (Chapter 4, Sections 42–56); development of a new district (Chapter 5, Sections 57–62); urban renewal (Chapter 6, Sections 63–73); organization and sources of funding (Chapter 7, Sections 74–78); and annexes (Chapter 9, Sections 81–87).

In Taiwan, the urban planning commission is equivalent to the Town Planning Board in Hong Kong. Similar to Hong Kong, the membership of the Urban Planning Commission is controlled by the mayor. As stipulated in Section 74, the Ministry of Interior, all levels of government and offices of villages, towns and county-level cities should appoint urban planning commissions for the analysis and review of urban plans. The constitution of the urban planning commissions is defined by the Executive Yuan. The problems with the formation and operation of the urban planning commissions will be discussed in the following chapter.

In 1994, the Construction and Planning Administration within the Ministry of Interior had commissioned a study by the Urban Planning Institute (UPI) of the Republic of China for a comprehensive review of the existing urban planning system to 'improve the qualities of environmental planning, development, conservation and management, and to achieve a balanced development in urban and rural areas through a coherent hierarchy of plans, and rational and efficient implementation institutions' (Ministry of Interior, 1994b, p. 3). The UPI has suggested a major reorganization of the administrative structure of planning through the establishment of a Ministry of National Planning which would coordinate and integrate plans at different levels (Fig. 14). It also recommended integrated planning for both urban and non-urban areas and the transformation of the regulatory development control system into a discretionary one. However, given the very complicated administrative structure and vested interests within and outside the bureaucracy, whether the study will lead to any fundamental changes in the planning system remains to be seen.

The following section examines the plan-making procedures at the municipal level in the three cases.

6.3. GOVERNMENT-LED PLAN MAKING PROCESSES AT THE MUNICIPALITY LEVEL

As discussed in Chapter 5, not all land use plans are statutory in the three political economies. The strategic plans in Hong Kong (TDSR, sub-regional development strategies) and the Concept Plan in Singapore are not statutory in nature. The National Comprehensive Development Plan for Taiwan is also prepared administratively by the CEPD at the central level. Although regional plans in Taiwan are prepared under the Regional Planning Law, citizen inputs are limited. In other words, visionary long-term plans in the three cases are prepared by the government. As there are limited inputs by the general public, one may argue that the strategic plans could only represent the visions of those in power.* However, a detailed discussion on strategic planning in the three cases merits another in-depth study† and therefore, the following focuses on the plan making process of local plans. Although the laws provide for citizen input in the making of urban land use plans in the three cases,

*As discussed, in recent years, the government of Hong Kong has developed the practice of consulting the public in preparing strategic plans.

†See Choo (1993), Ng (1993), Ng (1997), Planning Department (1996a), Urban Redevelopment Authority (1991).

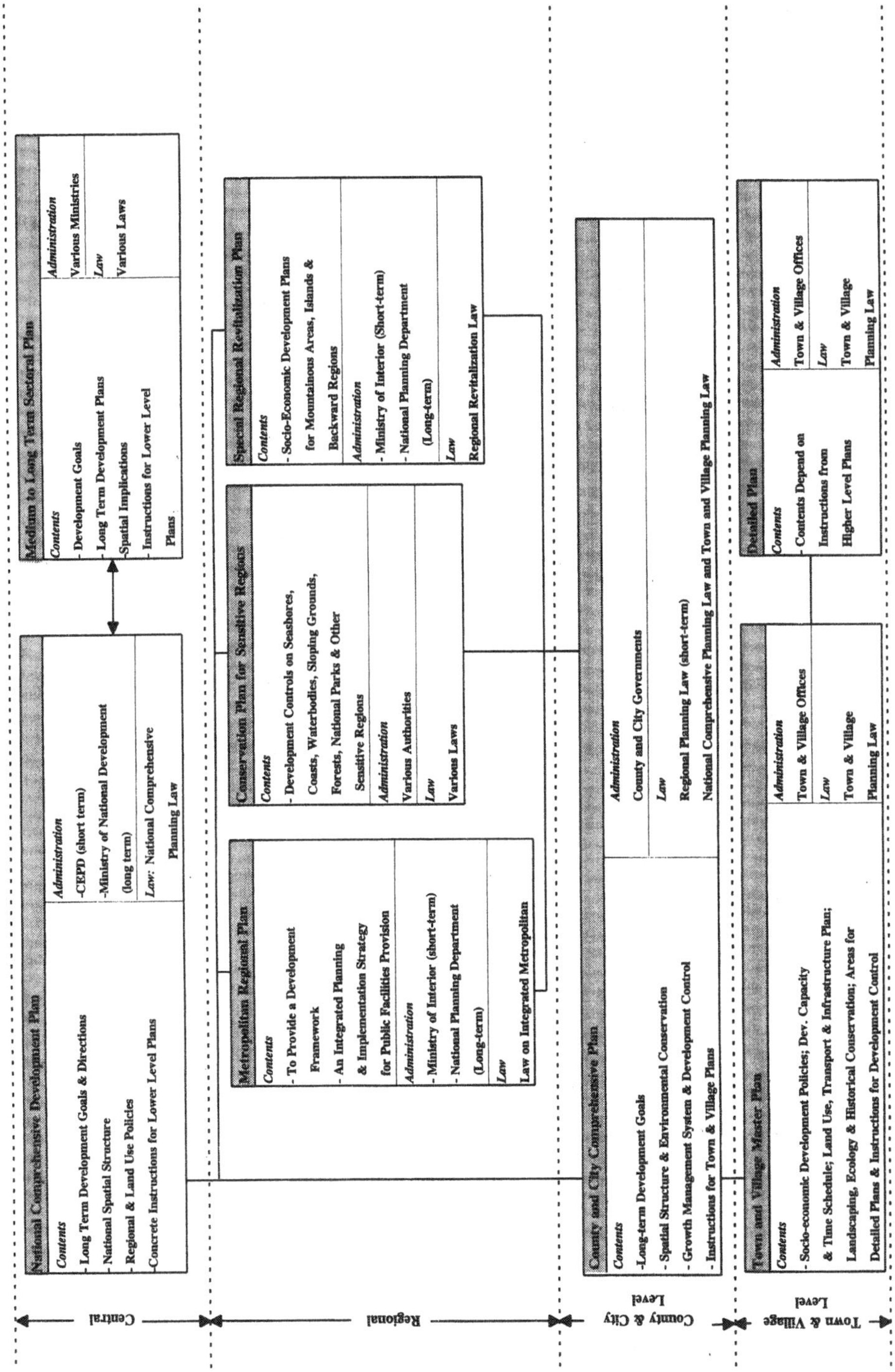

FIG. 14. Proposed land use administration, hierarchy of plans and related planning legislation in Taiwan. (Source: Ministry of Interior, 1994b, pp. 15, 23)

final decisions are made by the government. When non-government institutions are involved, such as the Town Planning Board in Hong Kong or the Urban Planning Commission in Taiwan, they are appointed bodies with little input from the lay public.

Table 6 below summarizes the steps involved in the plan making process within the three political economies.

6.3.1. Hong Kong

As provided in Section 3(1) of the TPO, the Town Planning Board shall undertake the systematic preparation of:

(a) draft plans for the lay-out of such areas of Hong Kong as the Chief Executive may direct, as well as for the types of building suitable for erection therein; and

(b) draft development permission area plans of such areas of Hong Kong as the Chief Executive may direct.

According to Sections 4 and 20(4), the above draft plans are basically land use zoning plans demarcating land for various uses, including comprehensive development areas where no specific zoning is made and planning permission by the Board is required for development in these areas.

As stipulated in Section 5, draft plans prepared under the direction of the Board, which the Board deems suitable for publication, shall be exhibited by the Board for public inspection at reasonable hours for a period of two months. It is also provided in Section 7(2) that amendments to a draft plan shall be exhibited by the Board for public inspection at reasonable hours for a period of three weeks. However, only

Table 6. Plan making processes in Hong Kong, Singapore and Taiwan

Steps involved	Hong Kong	Singapore	Taiwan
Identifying needs	The Government	The Minister for National Development and the competent authority appointed by the Minister	Government at all levels
Data collection/ preparation of plan and plan formulation	**Town Planning Board and the Planning Department**	The competent authority	Government at the appropriate level
Plan evaluation	**Government departments, the general public, and the Town Planning Board**	**The general public and the competent authority**	**The general public and the government at the appropriate level**
Plan recommendation	**The Town Planning Board**	The competent authority	Government at the appropriate level
Decision taking	The Chief Executive in Council	The Minister for National Development	**The Urban Planning Commission and then the government at higher level**

Bold: Involving non-government organizations or individuals.Source: Author.

those 'affected' by the draft plan so exhibited or an amendment to a draft plan may send a statement of their objections to the Board (Sections 6(1) and 7(4)).

Hearings are provided for the objectors to a draft plan or an amendment to a draft plan (Section 6). It is stipulated in Section 8 that '[a]fter consideration of all objections, the Board shall submit the draft plan, with or without amendments, to the Chief Executive in Council for approval. The Chief Executive in Council may approve, refuse to approve, or refer the draft plan to the Board for further consideration or amendment' (Section 9(1)(c)). Figure 15 shows the existing statutory plan making process in Hong Kong.

In the Town Planning White Bill, the government has proposed to enhance citizen participation in the plan making process. Figure 16 shows the proposed plan making process. 'To enable early public involvement in the plan-making process, the Bill provides for the publication of a planning study for public inspection and comment for a period of three months before a new outline zoning plan is prepared or major amendment is made to an existing plan' (Planning Department, 1996b, p. 5). The Bill also proposes to extend the publication period of an amended plan from 2–3 weeks to 2 months. Written representation in support of, or in opposition to, the provisions in the draft plan can then be made by any person (not only those who are affected by the plan) (Planning Department, 1996b, p. 6). The representations will be published for one month for comments and, after this, the Town Planning Board may hold a group inquiry to hear all the representations and comments. If these changes are passed, they would bring Hong Kong's plan making processes more in line with those in Singapore and Taiwan.

6.3.2. Singapore

As discussed previously, only the Master Plan is statutory in nature. As stated in Section 2 of the Planning Act, "Master Plan' means the Master Plan including the written statement submitted to and approved by the Governor in Council on 5th August 1958 under the provisions of Part IV of the Singapore Improvement Ordinance, repealed by this Act, and the rules made thereunder and includes all alterations and additions thereto submitted to and approved by the Governor in Council or the Minister under those provisions and rules and the provisions of Part II of this Act and the rules made thereunder.'

In Part II of the Planning Act, section 8 stipulates that the Minister may make rules to provide for the form, the content and the procedures to be followed in connection with the preparation, submission and approval of additions or alterations to the Master Plan. Section 6(2) provides that the competent authority may at any time submit to the Minister proposals for any alteration or addition to the Master Plan. In any case, the Master Plan should be reviewed at least once in every 5 years (Section 6(a)).

Section 8(a) specifies that 'the competent authority shall publish a notice of the submission to the Minister of any proposal for any addition or alteration to the Master Plan, and of the place or places where copies of the Master Plan and of the

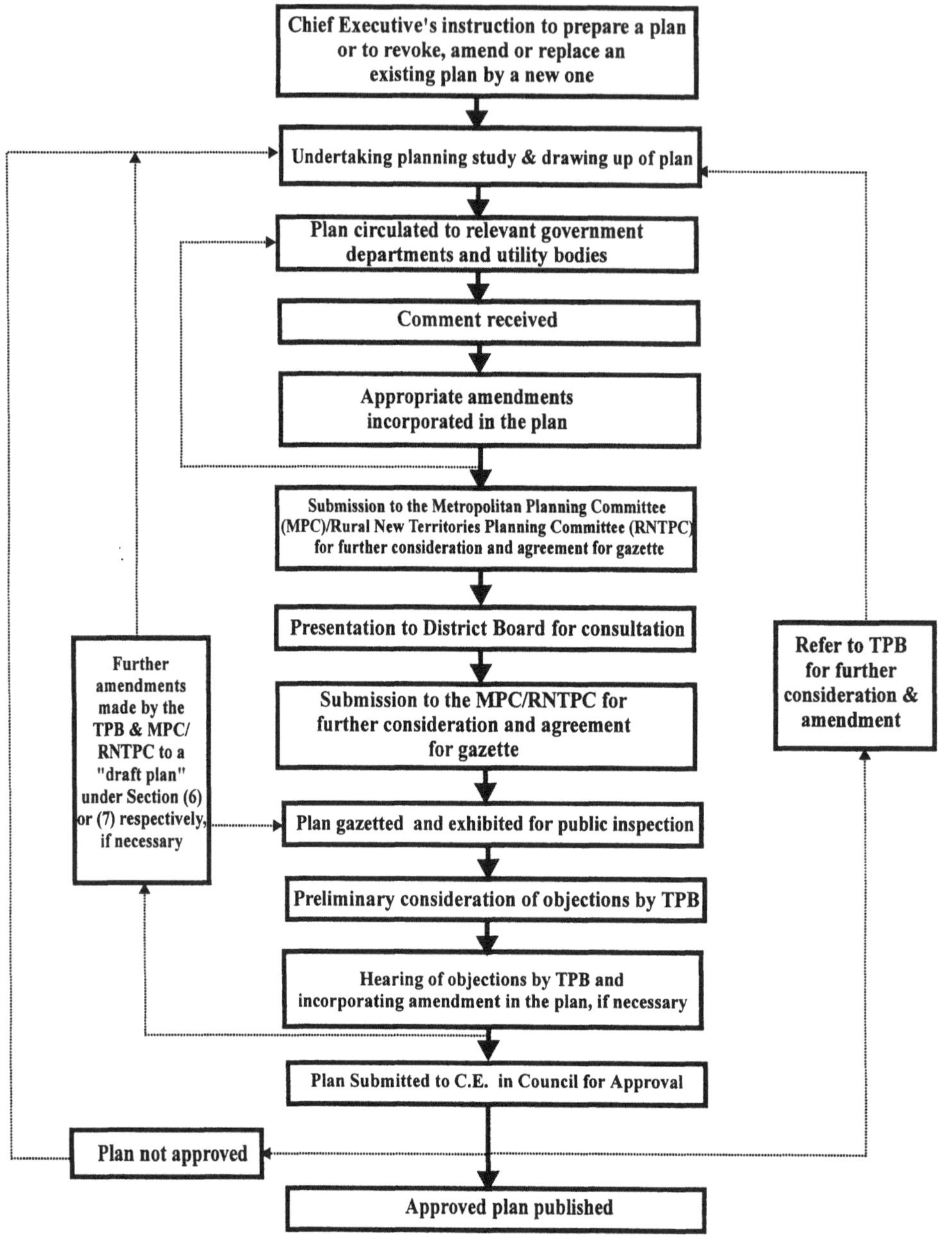

FIG. 15. Statutory plan-making process in Hong Kong. (Source: Planning Department, 1995, p. 22)

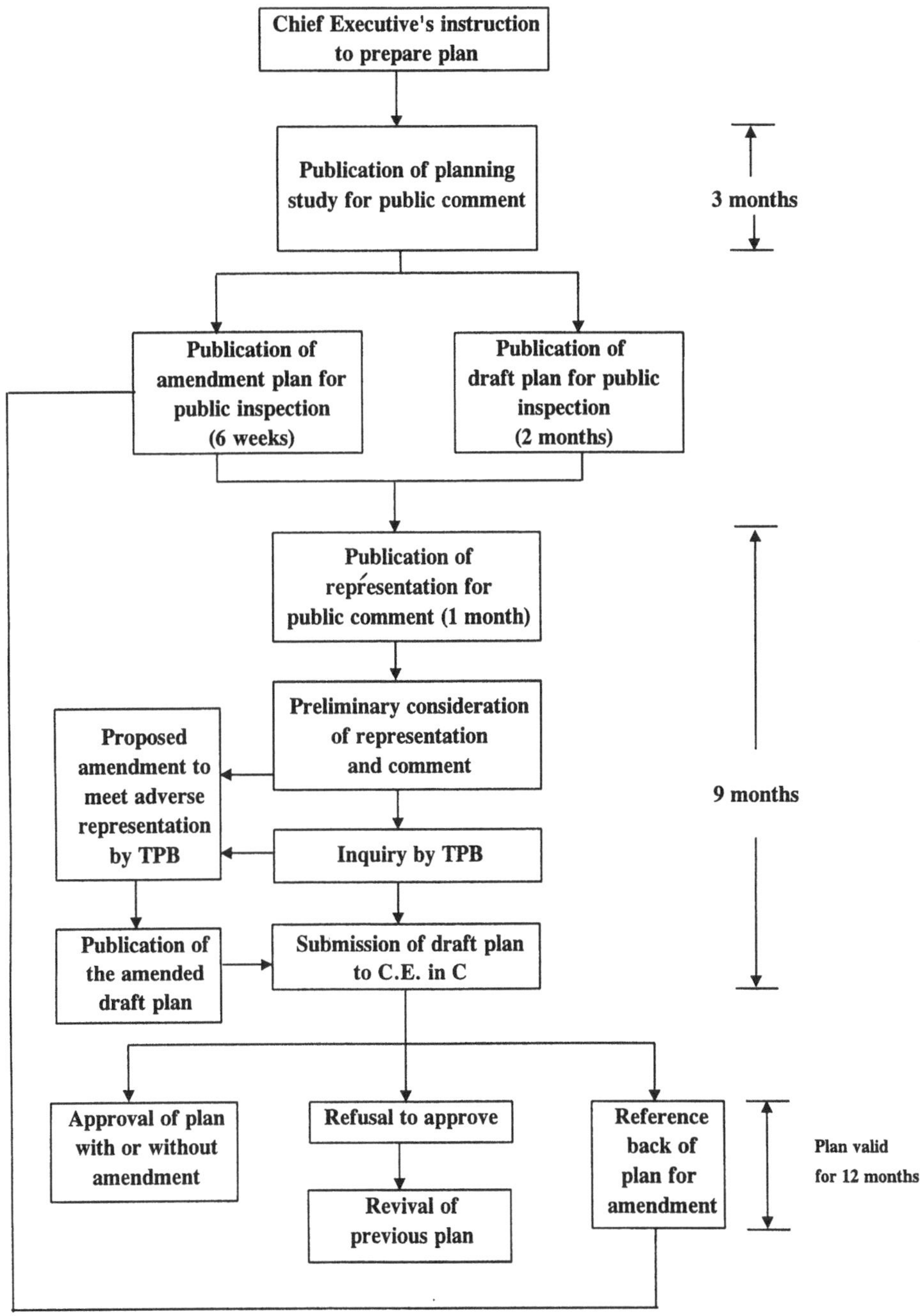

FIG. 16. The proposed plan-making process in Hong Kong. (Source: Planning Department, 1996b, p. 17)

proposal may be inspected.' In Section 8(b), it is stipulated that 'objections and representations duly made in accordance with the rules shall be considered, and that such public inquiries or other hearings as may be prescribed by the rules shall be held, before the Master Plan is added to or altered by the Minister.'

According to *The Master Plan Written Statement 1993*, with the completion of the long-range strategic *Concept Plan* in 1991, URA has proceeded to prepare Development Guide Plans (DGPs) for gazetting as the new Master Plan. The stages in preparing the new Master Plan are shown in Fig. 17. It should be noted that besides the provision of citizen participation in Section 8 of the Planning Act, outline DGPs of national or regional significance may also be exhibited or discussed in public dialogue sessions to solicit public feedback. In a way, this represents the trend of a more consultative government in Singapore as discussed in Chapter 3.

6.3.3. Taiwan

Unlike Hong Kong's Territorial Development Strategy and Singapore's Concept Plan, Master Plans for urban areas in Taiwan resemble a smaller scale zoning plan rather than a strategic plan. According to Section 5 of the Urban Planning Law, the Master Plan can be likened to a physical land use plan which specifies the boundary of the planning area, demographic trends and characteristics, spatial allocation of different land uses, transport network, social amenities, utilities and conservation areas. The Master Plan should also indicate the implementation schedule, sources of funding, etc. Section 17 specifies that districts in the first phase of development shall prepare detailed plans within 2 years after the publication of the Master Plan and complete the construction of public facilities within 5 years after the publication of the detailed plans.

'Detailed plan' as provided in Section 22 presents further details to the Master Plan: delimiting the planning area, specifying population density and capacity, land-use zoning control, road system, public facilities, etc. Similar to the provision of Master Plan review in Singapore, Section 26 of the Urban Planning Law in Taiwan stipulates that 'the planning authorities should review the plan at least once in five years on the basis of development conditions and recommendations from the public.'

The planning processes involved in the making of master and detailed plans are the same and so will be discussed together. In Taiwan, there are three categories of urban plans: city (town) plans, country and town plans, and special district plans (Section 9) which are made by various levels of governments (Table 7). Each of these plans should consist of the master plan specifications and maps, and the detailed plan specifications and maps (Sections 15–16, 22). Figure 18 shows the procedures involved in preparing, amending, approving and implementing the master and detailed plans as stipulated in the Urban Planning Law. Figure 19 shows how the procedures are refined in the Taipei Municipality. As provided in Section 66, the procedures also apply to the preparation and approval of urban renewal plans.

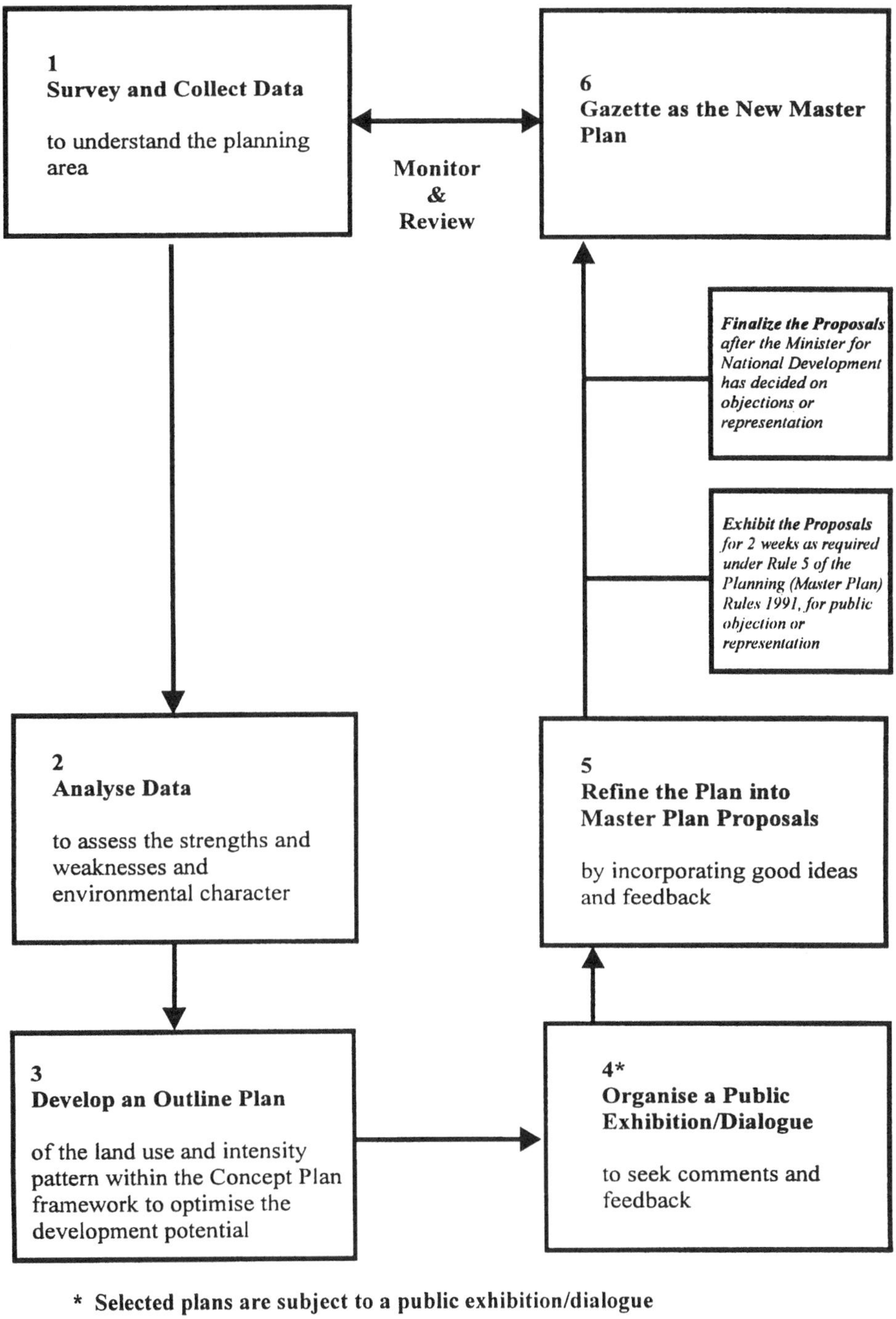

FIG. 17. The stages in preparing the New Master Plan in Singapore. (Source: Pamphlet issued by the Urban Redevelopment Authority, 1994)

Table 7. Preparation and approval of urban plans in Taiwan

Types of urban plans	Preparation by (Section 13)	Consideration by	Approval and record by (Section 20)
City (Town) Plans	Governments of municipalities, towns and county-level cities.[a]	*Section 19*:Draft plans will be exhibited for 30 days for public inspection. *Section 18*:Draft plans will be considered by the respective government or the Urban Planning Commissions of villages, towns and Provincial Cities. For plans prepared by higher level governments, relevant Provincial (Municipal) governments and offices of villages, towns or provincial cities shall first be consulted to solicit opinions for reference.	Capital City:approved by the Ministry of Interior;reported to the Executive Yuan for record Centrally Administered Municipalities, Provincial Capital and Cities: approved by the Ministry of Interior Seats of county Governments and County Level Cities: approved by the Provincial Government;reported to the Ministry of Interior for record
Town and Country Plans	Offices of towns, county level cities and villages or county Government.[b]		Approved by the Provincial Government and reported to the Ministry of Interior for record.
Special District Plan	Provincial (municipal) and County (city) Governments.[c]		Approved by the Ministry of Interior or the Provincial Government, and reported to the Executive Yuan or the Ministry of Interior for record.
Joint Urban Plan	By the relevant authorities in the adjacent administrative areas.[d]		Plans prepared by the Provincial Government shall be approved by the Ministry of Interior.

[a]For City (Town) Plan designated by the Ministry of Interior or the Provincial Government, when necessary, shall be prepared by the Provincial Government (section 14).[b]For Country and Town Plan designated by the Ministry of Interior or the Provincial Government, when necessary, shall be prepared by the Provincial Government (section 14).[c]Special District Plan, if necessary, shall be prepared by the Ministry of Interior (section 14).[d]If the areas do not go beyond the boundary of a province or county, the plan shall be prepared by the Provincial or County Government (section 13(3)).

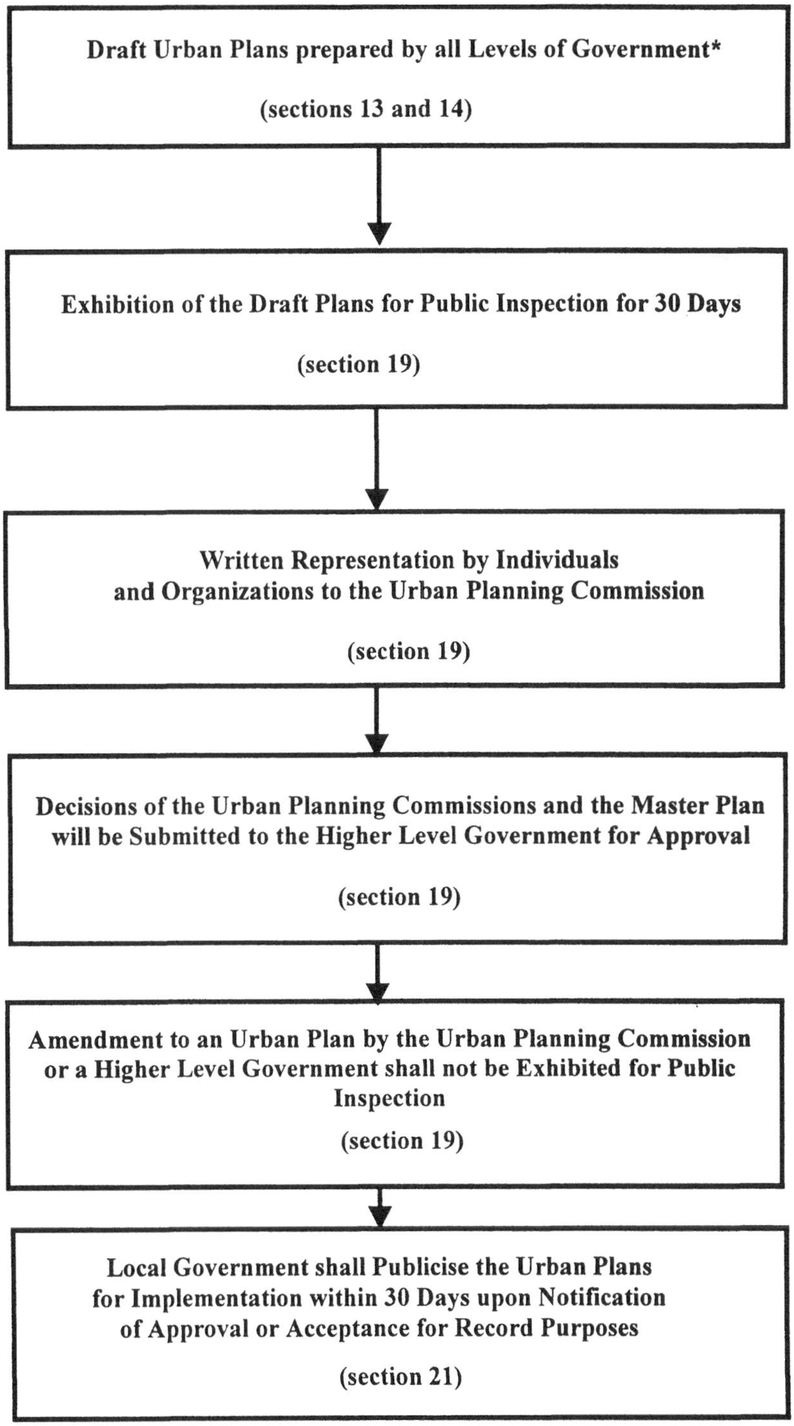

* Section 24 provides that "any person who has development right of a piece of land may, in conformity to the district zoning development plan, prepare his own detailed plan or amend a detailed plan, together with a project and financial plan. The application shall be submitted to local governments of centrally administered municipalities, counties (cities) or offices of village, towns and county-level cities to be processed" under section 23.

FIG. 18. Statutory plan making procedures in Taiwan

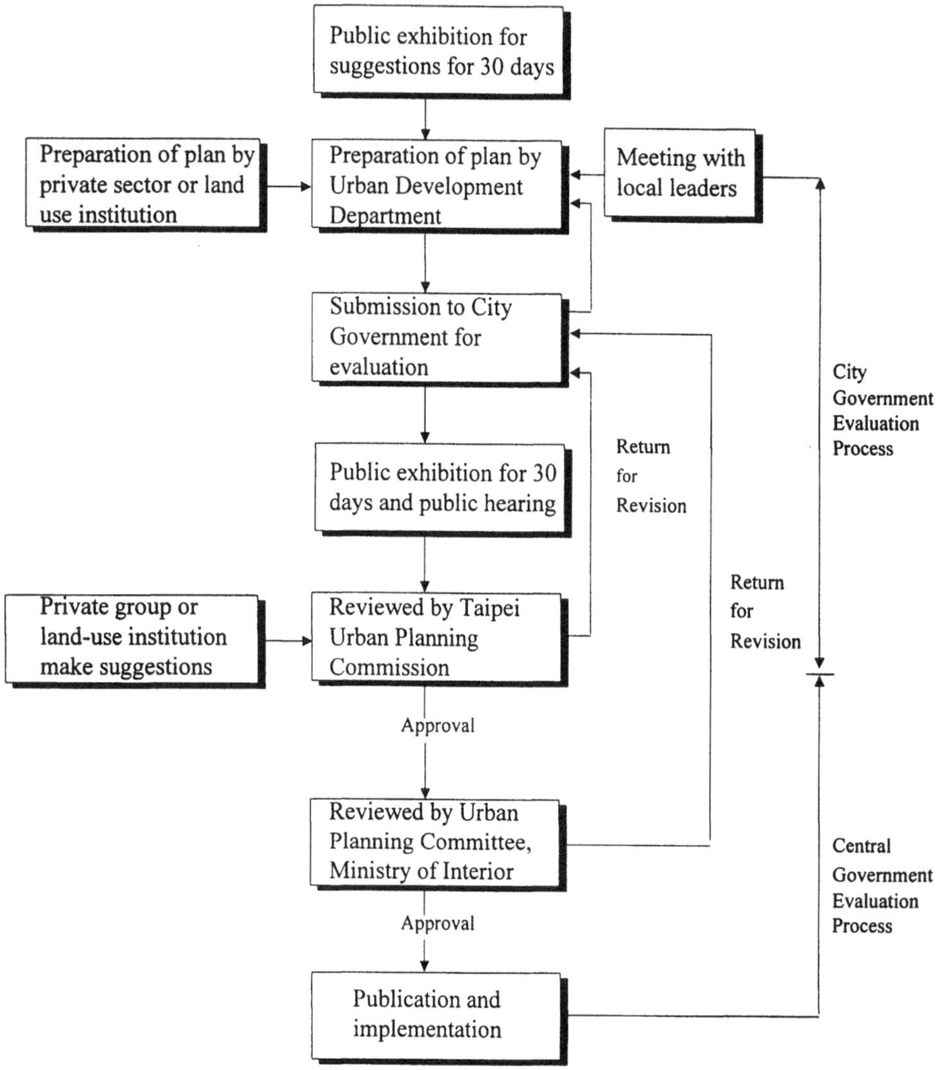

FIG. 19. City planning process in Taipei. (Source: Department of Urban Development, Taipei Municipal Government, 1994b)

Section 24 allows that 'any person who has development right of a piece of land may, in conformity to the district zoning development plan, prepare his own detailed plan or amend a detailed plan, together with a project and financial plan and submit them for consideration by the governments of centrally administered municipalities and counties (cities) or offices of villages, towns and county level cities'. However, since the implementation regulation* specifies that all land owners' consent is required before any person can prepare or amend a detailed plan, no one has used this provision to submit a detailed plan so far.

Section 74 in Chapter 7 of the City Planning Law which concerns organization and sources of funding, specifies that the Ministry of Interior, all levels of government and offices of villages, towns and county-level cities should appoint urban planning commissions for the analysis and review of urban plans. According to Section 19, before the plan is sent to the urban planning commission for consideration, the citizens can file written representations within the 30 days for public inspection. However, public inquiries and hearings, as may be held in the review of Singapore's Master Plan, are not provided in the Urban Planning Law in Taiwan. After considering the written representations, the decisions of the urban planning commission together with the master plan are sent to the higher level government for approval or amendment (Fig. 18). However, unlike the Hong Kongers or Singaporeans who may have a chance to comment on the revision and review of the statutory plans, the people in Taiwan are deprived of such a chance. As stipulated in Section 19, 'amendment to a master plan by the Urban Planning Commission or a higher level government shall not be exhibited for public inspection'.† As we shall see, this has serious implications for development control and plan implementation.

*Reference can be made to Section 6 of Chapter 2 of the 'Implementation Details of the Urban Planning Law in the Taipei Municipality,' *Urban Development Laws in the Taipei Municipality*, Department of Urban Development, the Taipei Municipality, p. 22 (in Chinese).

†In the City of Taipei, the Urban Planning Regulations allow another 30 days of public inspection after the plan is considered by the urban planning commission and examined by the local government.

Table 8. An extract of the notes attached to an outline zoning plan

RESIDENTIAL (GROUP C)	
Column 1: Uses always permitted	Column 2: Uses that may be permitted with or without conditions on application to Town Planning Board
Ancillary Car Park	Ambulance Depot
Flat	Bank
Government Staff Quarters	Clinic/Polyclinic
House	Education Institution
Police Reporting Centre	Fast Food Shop
Private Swimming Pool	Government Refuse Collection Point
Staff Quarters	Government Use (not elsewhere specified)
Utility Installation for Private Project	Hospital
	Hotel
	Petrol Filling Station
	Place of Recreation, Sports or Culture
	Private Club
	Public Car Park
	Public Transport Terminus or Station
Remarks: On land designated 'Residential (Group C)', any new building(s), and any addition, alteration and/or modification to the existing building(s) should not result in a total development or redevelopment in excess of a maximum building height as specified for each sub-area as set out below or the height of the existing building(s) whichever is the higher.	
Sub-area	Restriction
R(C)1	Maximum building height of 10.67 m
R(C)2	Maximum 12 storeys over 1 storey of carport

Source: Planning Department (1995, p. 14).

As stated in Section 16(1), where a draft plan or approved plan provides for the grant of permission for any purpose, an application for the grant of such permission shall be made to the Town Planning Board. Within 2 months of the receipt of the application, the Board may grant or refuse to grant the permission applied for (Section 16(3)). In considering a planning application, the Town Planning Board would usually consider factors such as planning intentions, government policies, social, economic and environmental impacts of the development on the wider area, traffic and infrastructure implications, compatibility of land uses (Planning Department, 1995, p. 27). Since TPB members are all appointed by the Chief Executive, who together with the Executive Council make final decisions to statutory plans, some people have queried the representativeness of the TPB in adjudicating development applications, especially when TPB makes decisions behind closed doors and the criteria for making decisions have not been made known to the general public. As TPB is serviced by the Planning Department, its independence has also been questioned.

If the applicant is aggrieved by the decision of TPB, the applicant may, within 21 days of being notified of the decision of the Board, apply in writing to the secretary of the Board for a review of the Board's decision (Section 17). This review takes place within 3 months of the receipt of the application and the applicant has 14 days notice of the time and place for a review, where the applicant or his authorized representative may attend and be given an opportunity to make representations (Sections 17(2) and 17(3)). If, following the review, the applicant is again dissatisfied by the TPB's decision, the applicant has the right to appeal to an independent Appeal

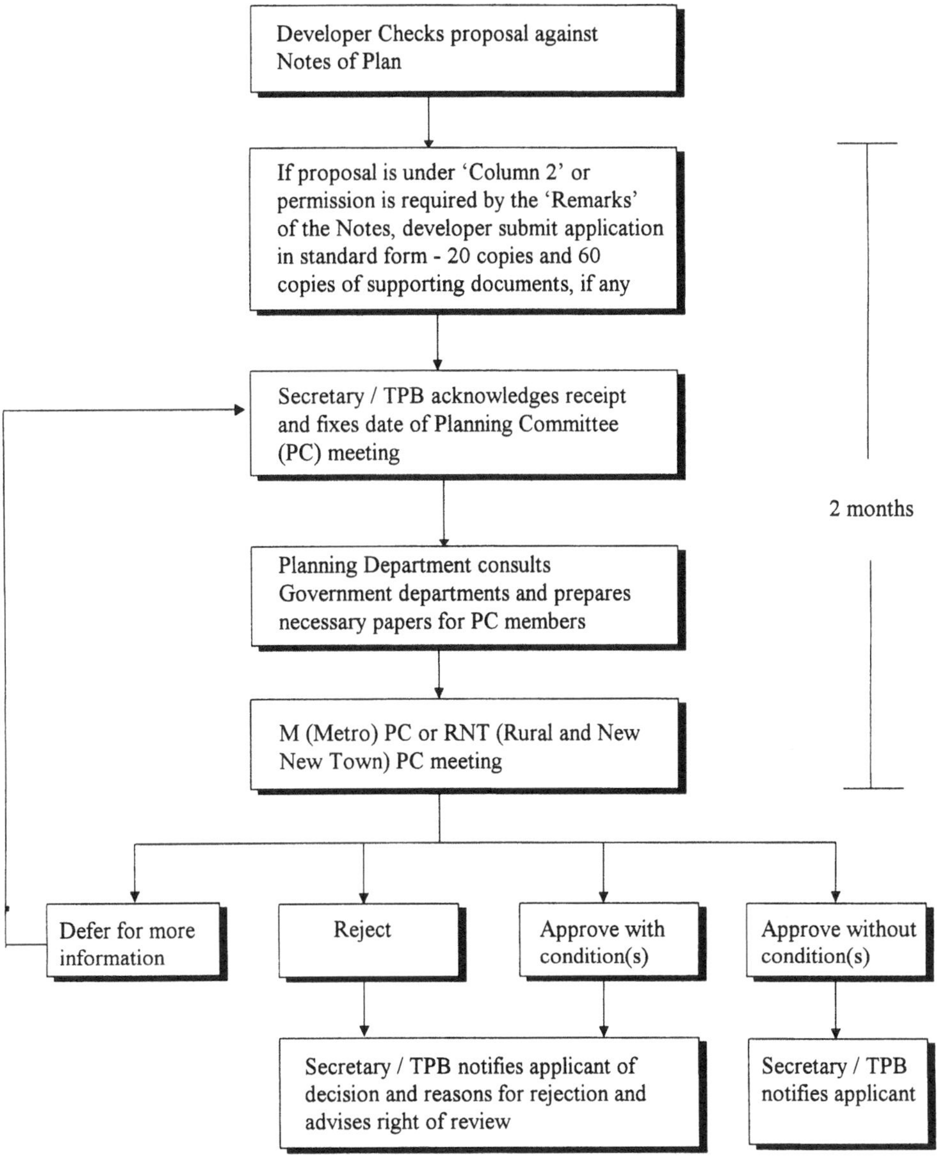

FIG. 20. Procedures for processing planning applications under Section 16 of the Town Planning Ordinance in Hong Kong. (Source: Planning Department, 1995, p. 28)

Board. The Appeal Board, although appointed by the Chief Executive, is independent in the sense that its members cannot be a member of the Town Planning Board or a public officer.

The Chief Executive in Council 'may make regulations prescribing the procedures to be followed in making an appeal, including matters to be set out in or to accompany the notice of appeal, the hearing of appeal and the determination of an appeal'

(Section 17B(4)). As stipulated in Section 17B(9), the decision of the Appeal Board on any appeal is final. However, the case may still be brought to the court for judicial review. Once brought to the court, the case will be considered from the point of law, not that of planning.

The Nam Sang Wai development project, related to one of the development permission area plans in the northwestern part of the New Territories, illustrates the system in operation. The project is controversial because the development proposed is located in the buffer zone of the Mai Po Marsh Conservation Area (a nature reserve of world-wide importance). The case is outlined below:*

- On 1 August 1992, Henderson Real Estate Agency Ltd applied for planning permission for a large scale residential development at Nam Sang Wai in the North West New Territories which lies south of Mai Po, a nature reserve of world-wide importance on the edge of the Inner Deep Bay. Surrounding the nature reserve on the landward side are two buffer zones and Nam Sang Wai lies on the southern edge of buffer zone 2.
- The development project proposed consist of 2550 residential units and an 18 hole golf course. Just over 1000 units were to be grouped in 25 blocks of seven or eight storeys. The population was expected to be about 9100.
- On 25 November 1992, the Town Planning Board rejected the application as the planning intention for the area was primarily 'to protect and conserve the landscape and ecological value of the area, and its scenic quality necessary to sustain Mai Po Nature Reserve' as set out in TPB guidelines, which were issued after Henderson's development application.
- Henderson appealed to the Town Planning Appeal Board. In March 1994, the Appeal Board allowed Henderson's appeal by a majority of 3 to 1 and planning permission was granted subject to certain conditions.
- The Town Planning Board applied for a judicial review and argued that the decision of the Appeal Board was wrong in law, irrational and ultra vires, since it contravened the planning intention for the area in question. The application for judicial review was dismissed on 28 April 1995. The TPB then appealed to the Court of Appeal, which allowed the appeal and quashed the decision of the Appeal Board.
- Henderson appealed to the Privy Council and the decision of the Appeal Board was restored in December 1996.

This case will definitely set a precedent for large scale residential development in the neighbouring areas of the Mai Po natural reserve. It also highlights various issues in the planning and development control mechanisms in Hong Kong. The planning intention of the statutory Development Permission Area Plan was not explicit. If the land had been zoned as a conservation area, Henderson Ltd would not have filed in a development application. When the TPB issued further guidelines, these were non-statutory documents and were not available to the lay public. This case highlights

---

*Summarized from *Privy Council Appeal No.54/96*, 'Decision of Privy Council on the Residential Development in Nam Sang Wai'.

the problem of closed door decision-making on issues of great public interest. It also illustrates the problem of discretionary planning systems which are not open to public scrutiny. Similarly, the constitution and operation of the Appeal Board also cause concern, as again the criteria for making decisions are not publicly known. Furthermore, development applications under the existing TPO are matters between the applicant and TPB with the result that even when the development project may be of territorial-wide significance, the general public have to rely on the good judgement and competence of TPB to make a decision. An analysis of the roles played by the various actors in this case will probably make a very interesting and important paper in the history of land use planning in Hong Kong. However, such an undertaking is beyond the scope of this paper.

Before the amendment of the Town Planning Ordinance in 1991, there was no provision for direct enforcement against breaches of planning control under the Town Planning Ordinance (Planning Department, 1995, p. 29). The 1991 Town Planning (Amendment) Ordinance introduced enforcement provisions for development permission area plans in the New Territories. The Ordinance defines 'unauthorized development' as development which is not an existing use, is not permitted under the development permission area plan and planning permission for the development has not been obtained (Planning Department, 1995, p. 29). Under the Ordinance, any person who undertakes or continues an unauthorized development commits an offence and is liable to a fine of HK$500 000 for the first conviction and HK$1 000 000 for second and subsequent convictions (Planning Department, 1995, p. 29).

Section 23(1) specifies that 'where there is or was unauthorized development, the Authority may, in a notice served on one or more of a land owner, an occupier or a person who is responsible for the unauthorized development

(a) specify the matters that constitute or constituted the unauthorized development; and

(b) specify a date by which if the unauthorized development has not been discontinued, the Authority requires

(i) it to be discontinued; or

(ii) permission for the development to be obtained under section 16.'

If permission for the unauthorized development has been refused and 'all rights of review or appeal under section 17 have been exhausted, abandoned or expired,' the Authority (Director of Planning) may require such a person 'to reinstate the land, by a date not earlier than 30 days after service of notice, to the condition it was in immediately before the development permission area became effective or to such other condition, more favourable to the person served, as the Authority considers satisfactory' (Section 23(3)). In specific cases where the unauthorized use constitutes a health or safety hazard, the Authority may serve a 'Stop Notice' requiring immediate discontinuance of the unauthorized development, and requiring steps to be undertaken to prevent such adverse effects (Planning Department, 1995, p. 29). Any noncompliance with these notices by a person is treated as an offence and the person is liable to a fine of HK$500 000 and a daily fine of HK$50 000. The fine for the second and subsequent convictions is double that for the first conviction.

Section 23(7) also provides that 'the Authority may enter the land and take whatever steps he considers necessary to ensure the discontinuance of the unauthorized development' (Section 23(7)). Section 22 provides the planning authority with the power to inspect the problematic sites. It is also stipulated in Section 24 that 'any person who is aggrieved by a decision of the Planning Authority may, within 30 days under service of the notice under section 23(3), apply in writing to the Secretary for Planning, Environment and Lands for a review of the Authority's decision'.

It should be noted that even in the *Town Planning White Bill*, no proposal has been made to extend enforcement power of the Authority to the main urban areas. In fact, the government relies on the Buildings Ordinance and the Building (Planning) Regulations and other ordinances such as Country Parks Ordinance, environmental legislation, Hong Kong Airport (Control of Obstructions) Ordinances etc. to enforce development.

The above has outlined the formal statutory means for controlling development in Hong Kong. It should be noted that similar to Taiwan, developers may apply for re-zoning of land use in a statutory plan. Such applications are dealt with administratively as the current TPO has no provision for dealing with re-zoning applications. If the TPB considers the application favourably, the mechanisms to amend the statutory plans will be activated to re-zone the land use. In addition to the statutory development control, as almost all the land in Hong Kong is under leasehold, the government as the lessor exercises development control in land leases. Such controls include the specification of uses, site coverage, plot ratio, building height, non-building areas, design and disposition, parking and loading/unloading requirements, environmental protection requirements and the provision of social facilities. Other development control mechanisms include building density zoning and density control incorporated in the Hong Kong Planning Standards and Guidelines and the schedule of Areas of Special Control in the main urban areas.

7.1.2. A discretionary development control system in Singapore

The power that the Minister for National Development has in 'directing' and 'controlling' development echoes the commanding position of the PAP-dominated state in Singapore's development process. As set out in Sections 30(1) and 30(2), the Minister may make rules generally to give effect to the provisions of the Planning Act including:

1. the development of land;
2. the control of density of buildings on land;
3. the regulation of the height, design, appearance and siting of buildings;
4. the control of means of access to land or building;
5. the number of car parks and car parking spaces to be provided on any class of land or building;
6. the protection of ancient monuments and land and buildings of historic and architectural interest;

7. the conservation of buildings, premises or land;
8. the form and manner in which applications for permission to develop or subdivide any land or to carry out any works within a conservation area shall be made;
9. the manner in which the competent authority shall deal with applications for permission to develop or subdivide any land or to carry out any works within a conservation area;
10. the fees or charges to be paid for any matter or thing done by the competent authority under this Act or any rules made thereunder;
11. the payment of a deposit by any person applying for permission to develop or subdivide any land or to carry out any works within a conservation area and the circumstances under which such deposit may be forfeited by the competent authority; and
12. the manner in which appeals may be made and determined under Sections 16 and 17 and the information to be supplied by the competent authority in connection therewith.

An important mechanism for the government to control and guide private development is through the Sale of Sites Programme which was initiated in 1967. Land parcels acquired or owned by the government have been assembled and sold to the private sector through a tender system. The land is sold with a set of conditions tailored to achieve the government's planning objectives (Urban Redevelopment Authority, 1995, p. 5). The general objectives of the programme are:

1. to provide land for private development; and
2. to set the direction and quality of urban redevelopment.

As Singapore has adopted a discretionary development control system to private sector development, all applications for permission to develop or subdivide land and building* are made to the competent authority (Section 10(4)) (Fig. 21). It is stipulated that no person shall, without the written permission of the competent authority, develop or subdivide any land or building. As mentioned, the meaning of 'develop' is defined in Section 12 as the carrying out of 'any building, engineering, mining or other operations in, on, over or under land, or the making of any material change in the use of any building or land'.

In considering the applications, Section 10(5) specifies that 'the competent authority shall act in conformity with the provisions of the Master Plan and any Certified Interpretation Plan in so far as they may be material thereto.' The Master Plan demarcated land into various land use zones which regulate the nature of their use as shown in the zoning and land/building use table in the Master Plan Written Statement (Table 9). The table shows the types of buildings or uses that may be permitted, not permitted, or permitted subject to special consideration in these zones.

---

*As specified in the interpretation section of the Planning Act, "land' includes buildings and any estate or interest in or right over land'. According to Khublall and Yuen (1991, p. 74), '[p]lanning control is extended to land and building subdivision to prevent indiscriminate subdivision that may result in fragmented ownership, substandard lot sizes and ultimately, problems in land use and development'.

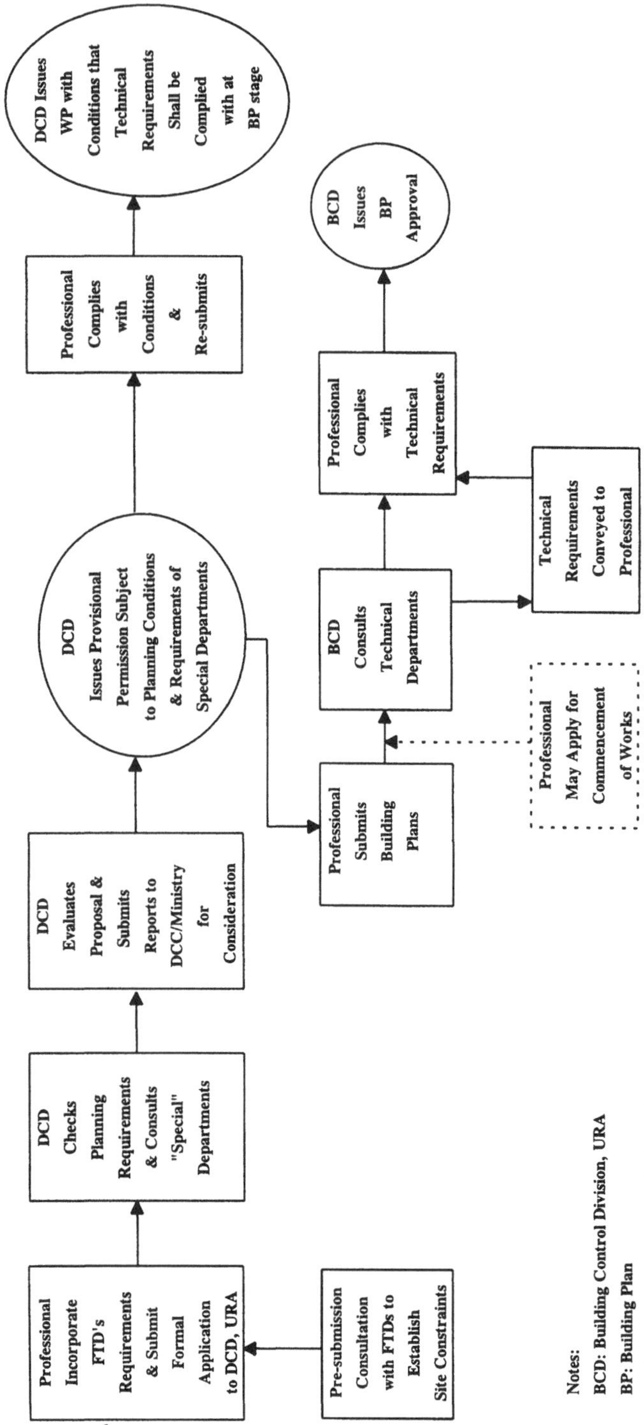

Notes:
BCD: Building Control Division, URA
BP: Building Plan
DCD: Development Control Division, URA
DCC: Development Control Committee
FTD: Front-line Technical Departments
WP: Written Permission

FIG. 21. Simplified planning approval system in Singapore. (Source: Khublall and Yuen, 1991, Fig. 4.1, p. 87)

Table 9. Zoning and land/building use table, Singapore

Land/Building Use \ Master Plan Zone	Residential	Residential with Commercial At 1st Sty only	Commercial/Residential	Commercial	Hotel	Clean/Light Industry	General/Special Industry	Warehouse	Business Park	Civic & Community Institution	Entertainment	Sports & Recreation	Hospital & Medical Care	Place of Worship	Education Institution	Transport Facilities	Agriculture
Detached House	P	C	C	X	C	C	X	X	C	C	C	C	X	X	X	X	C
Semi-detached House/Linked House	C	P	C	X	C	C	X	X	C	C	C	C	X	X	X	X	X
Terrace House	C	P	C	X	C	C	X	X	C	C	C	C	X	X	X	X	X
Cluster Housing/Townhouse	C	P	P	C	P	C	X	X	C	C	C	C	C	C	C	X	X
Residential Flat/Condominium	C	P	P	C	P	C	X	X	C	C	C	C	C	C	C	X	X
Service Apartment, Residential Club, Hostel	C	P	P	P	P	C	X	X	C	C	C	C	X	X	X	X	C
Boarding & Guest House	C	C	C	P	P	C	X	X	X	C	C	C	X	X	X	X	X
Shop	C	C	C	P	C	C	X	C	C	C	C	C	C	C	X	C	C
Tyre & Battery and Motor-cycle Repair	X	C	C	C	X	C	P	C	X	X	X	X	X	X	X	C	C
Restaurant	C	P	P	P	C	C	X	C	C	C	C	C	C	X	C	C	C
Foodshop	C	P	P	P	P	C	X	C	C	C	P	C	C	X	C	C	C
Market	C	C	P	P	C	C	X	C	C	X	C	C	X	X	X	C	C
Laundry and Dry Cleaner	C	P	P	P	C	P	P	C	C	C	X	C	X	X	C	X	C
Office	X	C	P	P	C	C	C	C	C	C	C	C	C	C	C	C	X
Hotel	X	C	C	P	P	C	X	X	X	C	C	C	X	X	X	X	X
Theatre, Cinema and Music Hall	X	C	C	P	C	X	X	X	C	C	P	C	X	X	C	X	X
Hospital and Related Institution	C	C	C	P	C	C	X	X	C	C	X	C	P	C	C	X	C
Educational	C	P	P	P	C	C	X	X	C	P	C	C	C	C	P	X	C
Non-Residential Club	X	C	P	P	C	C	C	C	C	C	P	C	C	C	C	X	X
Civic, Cultural and Community	C	C	P	P	C	C	X	X	C	P	C	C	C	C	C	X	C
Religious	C	C	C	X	X	C	C	C	C	C	X	X	C	P	C	X	C
Amusement Park	X	C	C	C	C	C	X	X	X	C	P	C	X	X	X	X	X
Sports and Recreation	C	C	C	C	C	C	X	X	C	C	C	P	C	X	C	X	C
Clean Industrial	C	C	C	C	X	P	C	C	P	X	C	C	X	X	X	X	C
Light Industrial	X	X	C	C	X	P	C	C	C	X	X	X	X	X	X	X	C
General Industrial	X	X	X	X	X	C	P	C	X	X	X	X	X	X	X	X	C
Special Industrial	X	X	X	X	X	X	C	C	X	X	X	X	X	X	X	X	X
Warehouse	X	X	X	X	X	C	P	P	C	X	X	X	X	X	X	X	X
Petrol Kiosk/Station	X	C	C	C	X	C	C	C	C	X	X	X	X	X	X	C	X
Car Park	P	P	P	P	P	P	P	P	P	P	P	P	P	P	P	P	C
Public Utilities	C	C	P	P	C	P	P	P	C	C	C	C	C	C	C	C	C
Agriculture	X	X	X	X	X	X	X	X	X	X	X	X	X	X	X	X	P

Notation: 'P' - Purpose for which buildings and lands may be used.
'C' - Purpose for which buildings and lands may be used subject to special consideration.
'X' - Purpose for which buildings and lands may not be used.

Source: Urban Redevelopment Authority (1993), *Draft Master Plan Written Statement*, Table 2, p.14

In practice, evaluation of a planning application is also guided by the non-statutory plans and planning policies and guidelines (circulars issued to professional bodies), site decision history, land safeguarding records and consultation with other government bodies. The competent authority, assisted by the Development Control Committee in the Urban Redevelopment Authority, may approve 'either unconditionally or subject to such conditions as he considers fit, or may refuse permission to develop or subdivide the land' (Section 10(5)). The Development Control Committee, unlike the Town Planning Board which has full decision making power, is an appointed advisory body that meets weekly to consider and make recommendations on development applications submitted by the private sector.

To speed up processing and avoid wasting time in preparing an elaborate submission which might be rejected, Section 11(1) allows the competent authority or the Minister to 'provisionally grant permission to develop the land for such period, not exceeding 6 months' before the application is determined, or in the case of an outline planning application. Once the application is determined, the provisional permission shall cease to have effect (Section 11(3)).

Any person who fails to apply for planning permission before developing or subdividing any land and building 'shall be guilty of an offence and shall be liable on conviction to a fine not exceeding S$3,000 and, in the case of a continuing offence, to a further fine not exceeding S$100 for every day after the first day during which the offence continues after conviction' (Section 10(8). For those who fail 'to comply with any condition imposed by the competent authority under section 10(5) or (6) or by the Minister under subsection (13) or by the Minister on appeal under section 17 shall be guilty of an offence and shall be liable on conviction to a fine not exceeding S$3,000' (Section 10(9)). As specified in Section 10(10), the competent authority or the Minister may cancel the permission when the person failed to comply with the imposed conditions.

These penalties also apply to those persons who fail to apply for written permission to 'carry out works in conservation area' which means 'any decorative, painting, renovation or building works which may affect the character or appearance, whether external or internal, of any building, premises or land situated within a conservation area but does not include any works which involve the development of land' (Section 14). Under Sections 10(13) and 13(11), the Minister can 'call in' any applications to develop or subdivide land or 'carry out works in conservation areas' for decision by him and his decision is final. Furthermore, the same sections specify that '[t]he decision of the Minister on any such application shall not be challenged or questioned in any court and the Minister may, in giving his decision, impose such conditions as he thinks fit to the granting of permission by the competent authority.'

It should be noted that once the formal approval is granted, it will lapse if the development is not completed within two years (Lim, 1987, p. 12). However, the applicant can request for an extension of the written permission.

If the applicant (whether applying for development, subdivision of land and building, or carrying out of works in conservation areas) is aggrieved by the decision of the competent authority, the applicant may 'appeal to the Minister within 28 days of the date of the notification of the decision in the form and manner prescribed by rules made under section 30' (Section 17(1)). The decision of the Minister will be final (Section 17(3)). However, no appeal is allowed in the following cases (Khublall and Yuen, 1991, pp. 110–111):

1. failure to give a decision within the proper time (3 months) (Section 10(11));
2. a grant of planning permission;
3. a refusal to extend planning permission after the statutory lapse of permission at the expiry of two years from the date of the grant of planning permission; and
4. the called-in application decided by the Minister (Section 10(13)).

Although there is no appeal board equivalent body in Singapore, the Ministry since 1963 has offered, in some instances, the appellant and the competent authority the opportunity to present their cases to the planning appeals inspectors at an oral hearing (Khublall and Yuen, 1991, p. 112). At present, the hearing is conducted by three planning appeals inspectors, one of whom is appointed the Chairman to ensure that recommendations on the appeal are well considered and the possibility of bias minimized (Khublall and Yuen, 1991, p. 114).

In relation to enforcement, Section 16 provides the planning authority in Singapore with the power of enforcing Sections 10 (restrictions upon development or subdivision of land and building) and 13 (restriction upon work within a conservation area) of the Planning Act. Section 19 allows authorized persons to enter into and upon any land during reasonable hours in the day for the purpose of carrying out the provisions of the Planning Act. 'Any person who wilfully obstructs a person at any time in the exercise of his authority shall be guilty of an offence and shall be liable on conviction to a fine not exceeding S$200 or to imprisonment for a term not exceeding 3 months or to both' (Section 19(3)).

As set out in Section 32 of the Planning Act in Singapore, a development charge is levied on 'written permission, including amendments to the written permission, granted under section 10(1) which permits development of land in an area:

1. not in accordance with the purposes for which the area has been zoned in the Master Plan;
2. in excess of the plot ratio specified in the Master Plan in relation to that area;
3. in excess of the equivalent plot ratio of that area; or
4. of such a nature involving any change in the use of the land or any building as may be prescribed.'

Figure 22 shows the development charge administrative system. The development charge can be seen as a form of betterment tax to recoup a part of the windfall gain resulting from a grant of planning permission over and above the intensity and use prescribed in the Master Plan (Lim, 1987, p. 2). This is different from Hong Kong, where no betterment charge is levied on permitted development unless there is a consequent need for lease modification. In such cases, the land owner or the developer has to pay a premium for lease modification.

Again, as specified in Section 36(1), the Minister may make rules for:

1. the different rates and methods of calculation of development charge;

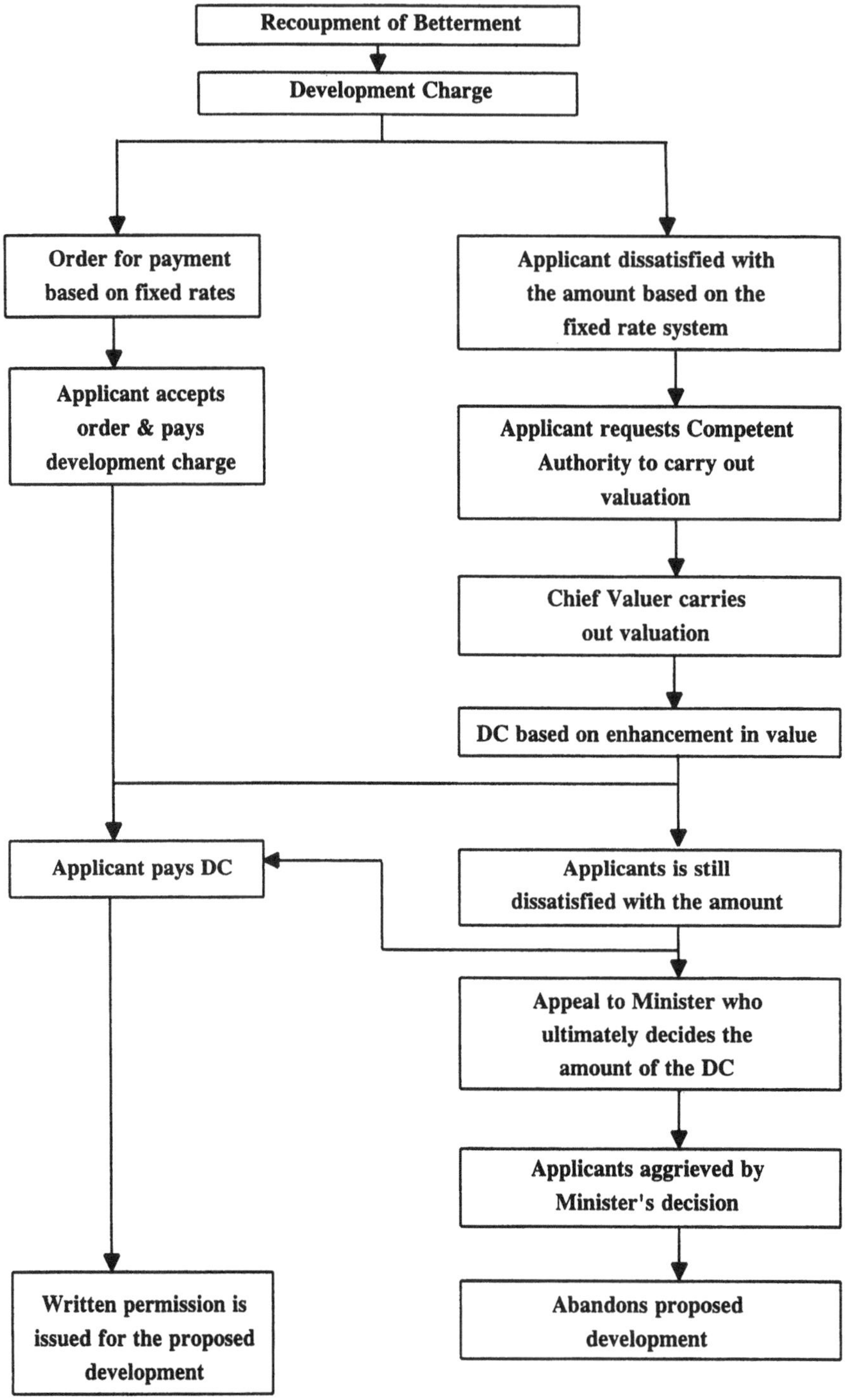

FIG. 22. Development charge administration system in Singapore. (Source: Khublall and Yuen, 1991, Fig. 11.1, p. 226).

2. exemption of a particular, or a class of, development from liability to pay the development charge;
3. the procedure for an application for the determination of any development charge; and
4. the refund, wholly or in part, of the development charge paid by any person.

Appeal is allowed if the applicant is dissatisfied with the decision of the competent authority on the development charge (Section 34(3)). The applicant may 'within 30 days of the service of the order, appeal to the Minister whose decision shall be final'.

7.1.3. A regulatory development control system in Taiwan

Unlike Singapore, where a discretionary planning permission system is adopted to guide private sector development, Taiwan adopts a regulatory development control system whereby land use zoning plans are used to guide private development. Most chapters in the Urban Planning Law in Taiwan concern plan preparation and implementation by the public sector, and in terms of controlling private development, the planners' role is more or less finished after the production or review of a 'rational' zoning plan. Planning approval is not required as long as the uses follow what have been zoned on the plan and specified in the law. So unlike Hong Kong and Singapore, where the developer or landlord can apply for development permission when the proposed development is different from that specified in the plan, this is not the case in the Urban Planning Law in Taiwan. Instead, the developer or the landlord may apply to the Urban Planning Commission for a change of land use. Amendment to a master plan by the Urban Planning Commission or a higher level of government does not need to be exhibited for public inspection (Section 19, Chapter 2). The Urban Planning Commission and the government can make changes to an approved master plan without informing the general public. Since members of all levels of the Urban Planning Commission are appointed by the Mayors, some people are willing to spend millions of dollars as a political investment to win the Nationalist Party nomination and election at local governments. Once they are in position, they could be in charge of, among others, the Urban Planning Commission and initiate land use changes to their benefit (Tien, 1989, p. 130). This has strengthened the 'gold-power' alliance considerably.

The amount of profit that a corporation or an individual could obtain from changing land uses can be illustrated by a case cited in Selya (1995, pp. 190–191). In 1987, the Cathay Insurance Company bought a plot of land for US$989.90 per square foot. It used the land as a parking lot to avoid the tax on unused land. Before the two-year grace period on developing the land as specified in city plans was to end, the company requested and received permission to change the land use designation from hotels to office buildings. This would allow the Company to sell the land at US$2277.80 a square foot. This type of speculative activity has contributed to escalating land prices.

Although Section 17 of Chapter 2 specifies that 'districts without detailed plans published should prohibit changes of building use and topographic changes,' it goes on to qualify this statement. 'Yet when the master plan has been published for more than two years, and the building line can be identified or major public facilities have been completed according to the master plan, the authorities responsible for building uses will define the building line and issue building permits'. This means that development can proceed in the absence of a detailed local plan as long as the master plan has been published for two years, which makes a rational monitoring of development very difficult.

Similar to Hong Kong and Singapore, there is no compensation for down-zoning in a plan. In Singapore, the Planning Act allows the government to levy a development charge (betterment tax) on projects which exceed the development limits specified in the plans. In Hong Kong, through the land lease system, a land premium maybe charged for upgrading of land uses in a renewal or modification of land lease. In Taiwan, in theory, any increments to land value due to increasing land costs should be recaptured by the government for the public good through the provisions of the Equalization of Urban Land Rights Statute (Selya, 1995, p. 140). The Statute specifies that land owners may declare the value of their holdings which the government compares with the prevailing values of land of similar class, location, and use. The owner and the government must then agree on the value of the property (Selya, 1995, p. 140). The Land Value Commissions at different levels of government are responsible for assessing the land value. It is reported that the property value finally agreed is usually a lot less than market rate so that those with vested interests in land can pay a lower tax (Chang, 1993b, p. 115). However, as compensation paid for land acquired for public purposes is pegged to this valuation exercise, the general public are usually not satisfied with the amount and have resisted land acquisition by the government. This practice has great implications when the government implements the plans (discussed in the following section).

The lack of development control is also reflected in the enforcement provisions in the Urban Planning Law. Unlike Hong Kong, enforcement provision is provided in all the urban areas in the Urban Planning Law. However, the implementation of this power by the government cannot be ascertained due to a lack of information. Given the general lack of respect for rules and regulations by the community, and the rather chaotic land use development in urban areas such as Taipei, one might conclude that enforcement actions are rarely exercised. This conclusion is supported by the serious fire hazards in restaurants within buildings not designed for commercial uses (Ministry of Interior, 1994b, p. 219). In fact, fires in restaurants, MTV and KTV studios have resulted in the deaths of several hundred Taipei residents yearly (Lee, R., 1993 quoted in Selya, 1995, p. 194). These are evidence that the government, entrepreneurs, developers and the public are not taking building rules and regulations seriously.

Another example is the Min Sheng New Community project where the government carried out a land pooling and replotting process and created some 118 ha of surplus land for infrastructure development and land sale (Selya, 1995, p. 146). The area was planned according to government regulations with above average public facilities and services provision. However, when the developers started construction, they increased the population capacity from 45 000 to 95 000. Infrastructure provision, of course, soon proved to be inadequate or non-existent and the general livability of the community declined. Some 14% of the land remained undeveloped (Selya, 1995, p. 147). The situation was worsened as the plans then had specified only the plot ratio but not the intensity of uses.*

Even when enforcement action is exercised, the penalty seems to be too light to be effective. Section 79 of Chapter 8 provides that 'for any construction, use of building, mining activity or topographical changes within the urban planning area which fail to comply with this Law or regulation made by all levels of government based on this Law, the local government or offices of villages, towns or county-level cities shall impose an order to demolish or reconstruct the building, discontinue the use or reinstate the land.' 'Any person who does not follow the order to demolish or reconstruct the building, discontinue the use or reinstate the land under section 79 commits an offence and will, in addition to have to carry out the order through administrative enforcement, be liable to a maximum of six-month imprisonment or hard labour, or a fine not exceeding NT$3,000.' The fine is obviously too light and it is doubtful if administrative enforcement is effective.

7.2. PLAN IMPLEMENTATION BY THE PUBLIC SECTOR

7.2.1. Hong Kong: fragmented responsibilities

Mechanisms employed by the government to implement plans have not been intensively researched. According to the Planning Department (1995, pp. 25, 26), the outline implementation programme prepared as part of the Territorial Development Strategy forms one of the bases for allocating public resources and determining priorities of major public works. At the district level, implementation of the planned land uses and infrastructure is guided by several types of programmes, such as public works development programme, new town development programme, public housing programme, community facilities programme and land sales programme, all under the control of various government departments. These programmes not only guide and control the timing of development of public sector projects, but also the disposal of land which in turn sets the pace for new private sector projects.

All land in Hong Kong is held by the government, which grants leasehold interests in accordance to the terms set out in Annex III to the Sino–British Joint Declaration (Government Information Services, 1996, p. 211). A land sales programme is issued at the beginning of each financial year and updated regularly, showing the details of public auctions and tenders normally held each month (Government Information Services, 1996, p. 211). Land is also made available at nominal premium to the Hong

*Since 1983, the Taipei Municipality has introduced limits other than the plot ratio in the 'Regulations on Land Use Zoning in the Taipei Municipality' (Department of Urban Development, Taipei Municipal Government, 1994a, pp. 33–112).

Kong Housing Authority for its public rental estates and Home Ownership Scheme, and to non-profit-making charitable, medical and educational institutions.

When private property is needed for the implementation of public works projects and cannot be acquired by negotiation, the use of compulsory powers becomes necessary. Property may then be acquired under ordinances that provide for payment of compensation, based on the value of the property, and for business loss, where appropriate, at the date of acquisition (Government Information Services, 1996, p. 211). It should be noted that this power has been used sparingly as such purchase is an expensive and slow process and subject to financial resources being available and approved by the government (Bristow, 1984, p. 162).

The Public Works Subcommittee under the Finance Committee of the Legislative Council is responsible for examining the administration's proposals for the upgrading of projects to, or downgrading from, Category A of the Public Works Programme, or changes to the scope and/or approved estimates of projects already in that category (Government Information Services, 1996, p. 8). The sub-committee makes recommendations on such proposals to the Finance Committee for approval.

7.2.2. Singapore: a collective responsibility

As discussed, statutory boards are important implementation agents for plans in Singapore. The most important ones include the Urban Redevelopment Authority, the Housing and Development Board, the Jurong Town Corporation, the Port of Singapore Authority and the Public Works Department. While the Master Plan is used to control private development, the Concept Plan is the means of guiding public investment in the built environment. The relevant public bodies are conferred with special statutory powers to implement their development programmes (Khublall and Yuen, 1991, p. 186).

The implementation of the broad proposals of the Concept Plan is effected by the Master Plan Committee (MPC). The MPC is chaired by the Chief Planner of the URA, with representatives from different government agencies such as the Ministry of National Development, the Ministry of the Environment, the Ministry of Trade and Industry, the Ministry of Defence, the Land Office, the Housing Development Board, the Public Works Department, the Economic Development Board and the Jurong Town Corporation (Koon, 1995, p. 121). Since the government owns at least 70% of the land in Singapore and is the largest developer of infrastructure, utilities, housing and industrial development, the MPC provides an important means of ensuring that land is developed according to the Concept Plan (Koon, 1995, p. 121). It also provides a channel for resolving conflicts amongst planning and implementation agencies (Koon, 1995, p. 121).

If private land is required for public development, the land may be acquired under the Land Acquisition Act. Unlike the case of Hong Kong where the power to resume land is used sparingly by the government, Section 5(1) of the Land Acquisition Act in Singapore states that land may be acquired:

1. for a public purpose;
2. for any work or an undertaking by any person, corporation or statutory board, which, in the opinion of the Minister, is of public benefit or of public utility or in the public interest; or
3. for any residential, commercial or industrial purposes.

Whereas in Hong Kong the payment of compensation is based on the value of the property at the date of acquisition, in Singapore, the date for the assessment of the market value of the land to be acquired is either:

1. 1 January 1986, if the acquisition is on or after 30 November 1987; or
2. 30 November 1973, for acquisition before 30 November 1987 (Khublall and Yuen, 1991, p. 199).

The market value on one of the above dates applies only if it is lower than the market value at the date of acquisition.*

In Singapore, the statutory boards are monitored by their respective ministries and by the Parliament. The types of control a minister can exercise over a particular statutory board are provided in the related acts or statutes and usually include providing policy directions; approving financial matters such as raising of loans from outside sources; and the appointment of chairman and members. The Parliament has to make sure that the financial and managerial policies followed by statutory boards are sound and helpful for promoting the aims of public policy underlying their establishments (Foo, 1992, p. 113).

7.2.3. Taiwan: a formidable task

The public sector is required by Sections 48, 58 and 68 of the Urban Planning Law to acquire land reserved for public facilities, development in new districts, and implementation of urban renewal plans through:

- resumption; or
- phased resumption; or
- urban land replanning (land pooling and replotting).

However, land resumption for these developments is hampered for a number of reasons. Similar to Hong Kong, Section 49 of the Urban Planning Law specifies that resumption of land reserved for public facilities should be compensated and the value is based on the average land value of neighbouring land not reserved for public facilities. If necessary, additional compensation is paid, the maximum of which not exceeding 40%. Section 50(1) provides that no income tax is collected from ad-

---

*In the case of residential properties, an *ex gratia* payment may be made to the Collector to bring the compensation up to the market value of the land as at the date of the acquisition, but this is subject to a ceiling of S$600 000 (Khublall and Yuen, 1991, p. 199).

ditional compensation of land reserved for public facilities. The provisions seem to be much more generous than the Land Acquisition Act in Singapore. However, as mentioned earlier, the average price set by the Land Value Commission has been suppressed so that the landowners can avoid paying high tax under the Land Law. As a result, the amount of compensation is not attractive and those whose properties are affected and need to be resumed tend not to cooperate with the local government at the implementation stage.

On the other hand, the duty of land resumption and construction of public facilities has become a very heavy financial burden to the resource hungry local governments. In 1984, only 16% of the land within the planned urban areas was allocated for the provision of public facilities (the standard should be 40%) and 63% of this allocated land was not resumed, developed and used (Chang, 1993a, p. 19). This is aggravated by the fragmentary ownership of public land. Although two-thirds of the land in Taiwan is publicly owned, 'sectionalism' exists and public authorities have been reluctant to hand over public land to local government for public land use development. Whereas section 42 of the Urban Planning Law provides that all public facilities should first utilize suitable public land, and section 59 stipulates that land managed by all levels of government should be handed over to local government for arrangement in accordance with the development plan for the new district, even public land held by public authorities may not comply with uses specified in the zoning plans.

Phased resumption is also not very effective in Taiwan.* In phased resumption, the government first resumes the land and then compensates the land owners. After land formation and infrastructure development, the former landlords have priority to buy back their land at cost. Like resumption, the governments need a considerable amount of capital to initiate phased resumption. Although the government could take 60% of the land resumed, in reality, two-thirds of the resumed land has to be used for infrastructure development. Therefore, the government could only retain one-third of land for sale. Moreover, the government has no control over subsequent developments and there is no time limit for land owners to build after the purchase of land.

Land pooling and replotting is another mechanism for the government to implement urban plans in Taiwan. The government helps the land owners to pool and replot their land. The land owners are responsible for all basic costs and infrastructure provision within the replotting area. The land owners enjoy tax exemption after they get back their due share of land in the replotting exercise and they enjoy tax reduction in subsequent transactions. Unlike the case of Singapore, where land is acquired by the government at the expense of the land owners, the benefits after phased resumption, pooling and replotting, and its consequent infrastructure development accrue to the land owners in Taiwan.

While the central government in Taiwan has been instrumental in building infrastructure projects to boost economic growth, local governments, unlike their counter-

*The discussions on phased resumption and pooling and replotting are based on Ministry of Interior (1994a, pp. 206–209).

parts in Hong Kong and Singapore which command considerable resources for plan implementation, face constant fiscal crises in implementing land use plans as stipulated in the Urban Planning Law.

CHAPTER 8

Conclusion

8.1. INTRODUCTION

This study set out to achieve three general goals of comparative studies: to account for existing planning mechanisms in the three cases; to improve existing practice through interpretation and transfer of foreign planning experience; and to advance planning theory. So far, the study has tried to provide an explanation for the existence of various land use planning mechanisms in the three economies. No explicit evaluation has been done to suggest experience that should be transferred among these three cases; nor have generalizations been made to advance theories transcending national cultural boundaries. These issues are discussed in the following sections, followed by a suggested research agenda for future studies.

8.2. FROM URBAN PLANNING MECHANISMS TO POLITICAL ECONOMY

The study argues that the political economy of a place will determine its nature and contents of urban planning mechanisms. While the term 'political economy' could embrace almost everything, this study focuses on how global economic changes and historical and cultural legacies affect the socio-economic and political milieu which shapes power relationships among key actors in a planning process, the development ideology, means of urban governance, substantial spatial concerns, and the operation of urban planning mechanisms. The diagnosis of urban planning mechanisms in the three cases reveals that the planning systems reflect the characteristics of the three respective political economies. This is most obvious in the case of Singapore.

The government-centred city state is able to maintain a strong top–down tightly controlled land use planning system to further the visions of its leaders. The leaders in Singapore are able to garner private resources to accomplish their plans for the general good. Viewed from a functional perspective, Singapore should be a planners' paradise as the dreams of planners as presented in physical plans are more likely to be materialized in the real world. However, in a political economy where a civil society 'hardly exists,' the question becomes: whose dreams are the plans representing? Singaporeans today can comment on local district plans (Development Guide Plans) which will eventually constitute the new statutory Master Plan. However, local plans are translations of broader policies, spatial or otherwise. If the citizens have little say in those broader issues, participating at the local plan making process will result in limited gain.

If Singapore's land use planning mechanisms are structured to ensure plan implementation and development control by the state, Taiwan's planning mechanisms are producing the opposite effects. While the administrative structure is very complicated and there are detailed procedures specified to produce spatial plans to 'guide' development, these have been neglected by the general lay public and exploited by the 'gold-power' alliance to further their own interests in land-related speculative activities. For the average citizens, a plan could mean a nightmare for them if their properties were down-zoned. In such cases, they would mobilize themselves and resist plan implementation. For the local governments, implementing the plans is almost impossible given the resource constraints. For progressive politicians, a good plan serves as a starting point to mobilize mass support. However, for land owners and developers, the plan is the guide for making a fortune, especially if they can succeed in re-zoning the land for more profitable activities. The politics of local land use planning reflects the broader political economy of Taiwan where the 'gold-power' alliance is taking advantage of the polity to further their interests which, however, are increasingly challenged by a democratizing political community. The story in Taiwan is, of course, complicated by the general lack of respect for rules and regulations as a social norm. Until people realize that a plan can be used as a means to further the general good, rather than the interests of the speculators, and take the plans seriously, the plans themselves are only marginally effective in guiding future development.

Hong Kong is a different story. The unbalanced power relationship between the government and the civil society is less dramatic than the case of Singapore. While the government and the private sector share common interests and have close links, such relationships, unlike those of Taiwan, are governed by a well established and respected legal system. With the democratization process, the government and the planners are trying a more open planning style. Although the strategic plans are non-statutory, the government has taken steps to consult the general public in the formulation of these. The publication of the *Town Planning White Bill* in 1996 has also allowed more citizen participation in the plan making and development process. Although the impacts which 'citizen voices' are having on the plans and the development process are difficult to ascertain, should this practice continue and consolidate, there is more hope of having plans incorporating the visions of the general public. However, whether this trend of planning will be checked by Hong Kong being a Special Administrative Region under Chinese rule remains to be seen.

8.3. TRANSFERABLE EXPERIENCE AND THEORETICAL IMPLICATIONS

A number of lessons with theoretical implications can be identified from these cases. As an activity in the public domain, land use planning needs to be seen as legitimate in the eyes of the various actors; otherwise, it is likely to be disregarded. The latter is the case of Taiwan where planning rules and regulations are not taken seriously by the public. Here urban planning mechanisms are not effective in providing a good physical environment. Legitimacy can be established through power, whether it is economic or political. However, when the global trend is to respect human rights and to espouse development which is sustainable rather than biased towards economic growth, consensus building may be a more effective means of establishing legitimacy. Consensus building in the planning and development process can ensure that a wide range of factors are considered before important decisions about the built environment are made. Solutions to the wicked urban problems and the vision for a desirable future need the contribution of every citizen. This is increasingly realized in democratizing Taiwan and Hong Kong, but seems to be a moot point in the case of Singapore.

The question, of course, is not choosing between a top–down or a bottom–up mode of planning. Striking a balance between participation and getting things done, and achieving an optimal mix of centralization and decentralization of power to plan and act are important agenda items in urban governance in the three cases. To a certain extent, Singapore has increased consultation with its population and Hong Kong has tried to reconcile participation and efficiency in planning and development. Taiwan, however, is a complicated case. On the one hand, Taiwan has suffered from a centralized public finance system which does not provide adequate resources for local authorities to implement their plans. Participation has limited meaning if local governments have little implementation ability. On the other hand, unlike the centrally controlled top–down planning system in Singapore, Taiwan suffers from a disjointed planning administrative structure with a proliferation of legislation and disconnected plans. Hence, the experience of Singapore having a central Ministry responsible for physical planning and development may serve as a useful lesson for Taiwan, recognizing that Singapore is a city state with just a single tier of government which makes its planning and development more manageable. The problems faced by Taiwan can serve useful lessons for Hong Kong as it is poised to integrate with the larger planning systems in China. Sooner or later, Hong Kong as a Special Administrative Region will face the challenge of reconciling not just global and local needs, but also national and regional (provincial) development pressures. A well coordinated administrative structure at different levels of government, producing integrated and related plans, provides important institutional infrastructure for effective planning.

Another lesson that can be derived from the cases is that the ability to command resources by local governments is essential for bottom–up planning initiatives and plan implementation. By this approach, the planning authority is in a better position to solicit the collective wisdom of the general public in the plan making and development process, and to utilize the energy of the private sector interests.

It is interesting to see that the three cases typify three major means of urban governance: Hong Kong with its emphasis on the market forces (economics); Singapore with its emphasis on the role of the state (planning); and Taiwan with its emphasis on power play within and among various actors (politics). As a result, the roles of land use planning differ in the three political economies. Future evolution of the roles of urban planning mechanisms in the course of urban development will depend on the dominant means of governance adopted, which in turn will depend on the power relationships amongst the major actors in the three cases.

8.4. FUTURE RESEARCH AGENDA

Comparative studies on urban planning mechanisms in Asia is a largely unexplored field and much further work is still required. This study is just a preliminary one to compare urban planning mechanisms within different political economies and more indepth and thorough research is required to verify various arguments. The following are some suggestions:

1. Histories of land use planning ideologies and administration.
2. Hierarchy of plans, their coherency and relationships.
3. Strategic planning processes and strategic plans.
4. Detailed case studies on development control and enforcement.
5. Roles of land use planners in urban governance.
6. Relationship between spatial and sectoral plans.
7. Relationship between international, national, regional and local planning.
8. Roles of various actors in the planning making and development process: private sector, community with diversified interests, professional planners, politicians, etc.
9. Planning legislation and its enforcement.
10. Financing the implementation of plans.
11. Informal planning practice through in-depth case studies.
12. Utilizing theories of urban development to understand planning practice in Asia and to distill from the studies theoretical implications for planning in the public domain.

Bibliography

ADB (1996) *Asia Development Outlook 1996 and 1997*. Oxford University Press (China), Hong Kong.

Akrasanee, N. and Chirathivat, S. (1989) Trade issues among ASEAN, Asian NICs and Japan. In *Global Adjustment and the Future of Asian-Pacific Economy*, eds Shinohara, M. and Lo, F. C. Asian and Pacific Development Centre and Institute of Developing Economies, Kuala Lumpur, Malaysia.

Amirahmadi, H. and Gladstone, D. (1996) Towards a dynamic theory of the state and civil society in the development process. *Journal of Planning Education and Research* **16**(1), 15–25.

Appelbaum, R. P. and Henderson, J. (1992) *States and Development in the Asian Pacific Rim*. Sage, London.

Ballantine, J. W. (1952) *Formosa. A Problem for United States Foreign Policy*. The Brookings Institution, Washington D.C.

Booth, P. (1996) *Controlling Development: Certainty and Discretion in Europe, the USA and Hong Kong*. UCL Press, London.

Bourne, L.S. (1984) Cross national comparative planning studies: commentary. *Town Planning Review* **55**(2), 150–151.

Bristow, R. (1984) *Land Use Planning in Hong Kong: History, Policies and Procedures*. Oxford University Press, Hong Kong.

CEPD (1987) *A Brief Introduction to the National Comprehensive Development Plan for Taiwan*. Urban and Housing Development Department, Council for Economic Planning and Development, Republic of China (in Chinese).

CEPD (1991) *Urban and Regional Development Statistics*. Urban and Housing Development Department, Council for Economic Planning and Development, Republic of China, Executive Yuan.

CEPD (1992) *Taiwan Statistical Data Book*. Council for Economic Planning and Development, Republic of China.

CEPD (1993) *Urban and Regional Statistics*. Urban and Housing Development Department, Council for Economic Planning and Development, December 1993.

CEPD (1995) *Taiwan Statistical Data Book*. Council for Economic Planning and Development, Republic of China, Table 3-13a.

Chang, K. S. (1993a) Local government, citizen organizations and local planning: the political economy of urban planning in the 1990s. In *Challenges of Territorial Planning*, Chang Yung-Fa Foundation, Institute for National Policy Research, pp. 157–176 (in Chinese).

Chang, K. S. (1993b) *Urban Planning in Taiwan (1895–1988)*. Chang Yung-Fa Foundation, Institute for National Policy Research (in Chinese).

Chen, H. C. (1974) Politics in an administrative state: where has the politics gone? Paper presented at *Seminar on Trends in Singapore*, 24 November 1974, sponsored by the Institute of Southeast Asian Studies, Singapore.

Chen, H. C. (1976) The political system and political change. In *Singapore: Society in Transition*, ed. Hassan, R. Oxford University Press, Kuala Lumpur, pp. 30–51.

Chen, P. S. J. (1983) Singapore's development strategies: a model for rapid growth. In *Singapore Development Policies and Trends*, ed. Chen, P. S. J. Oxford University Press, Singapore, pp. 3–25.

Chong, Wai Leung (1996) Power boom of the underworld triggered alarm clock in Taiwan's political arena. *Ming Pao*, 25 November, p. A10 (in Chinese).

Choo, M. L. (1993) Environmental planning. In *Action Proposals: Environmental Issues in Development and Conservation*, ed. Briffett, C. School of Building and Estate Management, National University of Singapore, SNP.

Chu, Y. H. (1994) The realignment of business-government relations and regime transition in Taiwan. In *Business and Government in Industrialising Asia*, ed. MacIntyre, A. Cornell University Press, Ithaca, New York, pp. 113–141.

Chua, B. H. (1996) Culturalisation of economy and politics in Singapore. In *Pathways to Asia: The Politics of Engagement*, ed. Robison, R. Allen and Unwin, Singapore, pp. 87–107.

Clad, J. (1989) *Behind the Myth: Business, Money and Power in Southeast Asia*. Unwin Hyman, London.

Cullingworth, J.B. (1993) *The Political Culture of Planning: American Land Use Planning in Comparative Perspective*. Routledge, New York.

Cuthbert, A.R. (1985) Architecture, society and space—The high-density question re-examined. *Progress in Planning* **24,** 71–160.

Cuthbert, A.R. (1991) For a few dollars more: urban planning and the legitimation process in Hong Kong. *International Journal of Urban and Regional Research* **15**(4), 575–593.

Davies, S.N.G. (1977) One brand of politics rekindled. *Hong Kong Law Journal* **7**(1), 44–48.

Department of Urban Development, Taipei Municipal Government (1994a) *Urban Development Laws in the Taipei Municipality*. Department of Urban Development, Taipei Municipal Government (in Chinese).

Department of Urban Development, Taipei Municipal Government (1994b) *International City for the 21st Century—My Home in Taipei*. Department of Urban Development, Taipei Municipal Government, Taipei, July.

Deyo, F. C. (1987) *The Political Economy of the New Asian Industrialism*. Oxford University Press, New York.

Douglass, M. (1994) The 'developmental state' and the newly industrialised economies of Asia. *Environment and Planning A* **26,** 543–566.

Executive Yuan (1991) *A Brief Introduction to the Work of Government Organizations in the Republic of China*. Executive Yuan, Taiwan.

Faludi, A. and Hamnett, S. (1975) *The Study of Comparative Planning*. Centre for Environmental Studies, London.

Far Eastern Economic Review (1995) *Asia 1995 Yearbook: A Review of the Events of 1994*. Review, Hong Kong.

Fei, J.C.H., Ranis, G. and Kuo, S.W.Y. (1979) *Growth with Equity: The Taiwan Case*. Oxford University Press, Oxford: published for the World Bank.

FEER (Weekly) *Far Eastern Economic Review*, Hong Kong.

Foo, A. F. (1992) Planning process: implementation, coordination and control in the city-state of Singapore. *School of Architecture Journal*, National University of Singapore, Singapore, pp. 103–118.

Friedmann, J. (1988) *Life Space and Economic Space: Essays in Third World Planning*. Transaction Books, New Brunswick.

Geiger, T. (1973) *Tales of Two City-States: The Development Progress of Hong Kong and Singapore*. National Planning Association, Washington D.C.

Government Information Services (1996) *Hong Kong 1996*. Government Printer, Hong Kong.

Heikkila, E.J. and Griffin, M. (1995) Confucian planning or planning confusion? *Journal of Planning Education and Research* **14,** 269–279.

Hobbs, P. (1992) The economic determinants of post-war British town planning. *Progress in Planning* **38,** 179–300.

Hong Kong (1994) *Chapter 131 Town Planning Ordinance*. Government Printer, Hong Kong.

Hong Kong Census and Statistics Department (1994) *Revised Estimates of GDP 1961 to First Quarter 1994*. Government Printer, Hong Kong.

Hsiao, M. H. H. (1981) *Government Agricultural Strategies in Taiwan and South Korea: A Macrosociological Assessment*. Institute of Ethnology, Monograph Series B, No. 9. Academia Sinica, Taiwan.

Hughes, H. (1988) *Achieving Industrialization in East Asia*. Cambridge University Press, Cambridge.

International Labour Organization (1981) *Export-led Industrialization and Development*. Koon Wah Printing, Singapore.

Johnson, W.C. (1984) Citizen participation in local planning in the U.K. and U.S.A.: A comparative study. *Progress in Planning* **21,** 149–221.

Jones, D.M. and Brown, D. (1994) Singapore and the myth of the liberalizing middle class. *The Pacific Review* **7**(1), 79–87.

Kerr, G.W. (1965) *Formosa Betrayed*. Houghton-Mifflin, Boston.

Khakee, A. (1992) *Cross-National Transfer of Planning Ideas between China and Sweden*. Swedish Council for Building Research, Stockholm.

Khakee, A. (1996) Urban planning in China and Sweden in a comparative perspective. *Progress in Planning* **46,** 91–140.

Khublall, N. and Yuen, B. (1991) *Development Control and Planning Law: Singapore*. Longman, Singapore.

Koon, H.C. (1995) Urban land use planning in Singapore: towards a tropical city of excellence. In *Environment and the City*, ed. Ooi, G. L. The Institute of Policy Studies, Times Academic Press, pp. 109–128.

Lee, K. (1993) *New East Asian Economic Development—Interacting Capitalism and Socialism*. M.E. Sharpe, Armonk.

Lee, R.F.F. (1993) Autonomous managing key in buildings rights. *Free China Journal* 5 March, 4.

Lim, C. Y. (1991) *Development and Underdevelopment*. Longman Singapore, Singapore.

Lim, L. and Pang, E. F. (1991) *Foreign Direct Investment and Industrialization in Malaysia, Singapore, Taiwan and Thailand*. OECD Development Centre, Paris.

Lim, L. Y. (1987) *Development Charge*. Singapore National Printers, Singapore.

Lo, F.C., Song, B.N. and Furukawa, S. (1989) Patterns of development and interdependence among the east and southeast Asian economies. In *Global Adjustment and the Future of Asian-Pacific Economy*, eds Shinohara, M. and Lo, F.C. Asian and Pacific Development Centre and Institute of Developing Economies, Kuala Lumpur, Malaysia.

MacIntyre, A. (1994) *Business and Government in Industrializing Asia*. Cornell University Press, Ithaca, New York.

Mann, J. (1994) Taiwan to U.S.—we're back! *The Los Angeles Times* 28 July.

Masser, I. (1984) Cross national comparative planning studies. *Town Planning Review* **55**(2), 137–160.

Ministry of Construction and Transport, South Korea (1996) *Proceedings of the First Conference on City and Housing Development in China and South Korea*. People's Republic of China, Beijing (in Korean and Chinese).

Ministry of Interior (1994a) *Research on the Comprehensive Review of the Urban Planning Law: Appendices*. Construction and Planning Administration, Ministry of Interior (in Chinese).

Ministry of Interior (1994b) *Research on the Comprehensive Review of the Urban Planning Law*. Construction and Planning Administration, Ministry of Interior (in Chinese).

Ng, M.K. (1993) Strategic planning in Hong Kong. *Third World Planning Review* **64**(3), 287–311.

Ng, M. K. (1995) Urban and regional planning. In *From Colony to SAR: Hong Kong's Challenges Ahead*, eds Cheng, J.Y.S. and Lo, S. The Chinese University Press, Hong Kong, pp. 227–260.

Ng, M. K. (1997) Strategic planning and the territorial development strategy review. *Hong Kong Public Administration*, 6(1, March), 31–59.

Ng, M.K. and Cook, A. (1997) Reclamation: an urban development strategy under fire. *Land Use Policy* **14**(1), 5–23.

Ng, M. K. and Xu, J. (1996) Urban planning in Greater China: A comparative study of Guangzhou, Hong Kong and Taipei. Paper presented at the *ACSP-AESOP Joint International Congress: 'Local Planning in a Global Environment'*, Toronto, Canada, 25–28 July.

Paul, E.C. (1993) Prospects for liberalization in Singapore. *Journal of Contemporary Asia* **23**(3), 291–305.

Perkins, D.H. (1994) There are at least three models of East Asian development. *World Development* **22**(4), 655–661.

Planning Department (1995) *Town Planning in Hong Kong: A Quick Reference*. Government Printer, Hong Kong.

Planning Department (1996a) *Consolidated Technical Report on the Territorial Development Strategy Review 1996*. Planning Department, Hong Kong Government.

Planning Department (1996b) *Consultation Paper on Town Planning Bill*. Government Printer, Hong Kong.

Przeworski, A. and Teune, H. (1970) *The Logic of Comparative Social Inquiry*. Wiley, New York.

Rodan, G. (1996) Singapore. In *The New Rich in Asia*, eds Robison, R. and Goodman, D. S. G. Routledge, London, pp. 20–45.

Selya, R. M. (1995) *Taipei*. John Wiley, Chichester.

Sewell, W. Jr. (1967) Marc Bloch and the logic of comparative history. *History and Theory* **6**, 208–218.

Shinohara, M. (1989) Global adjustment and the role of Japan. In *Global Adjustment and the Future of Asian-Pacific Economy*, eds Shinohara, M. and Lo, F. C. Asian and Pacific Development Centre and Institute of Developing Economies, Kuala Lumpur, Malaysia.

Singapore (1990) *Planning Act (Chapter 232)*. Government Printer, Singapore.

Singapore Department of Statistics (1983) *Economic and Social Statistics Singapore 1960–1982*. Department of Statistics, Singapore.

Singapore Department of Statistics (1991) *Yearbook of Statistics Singapore 1991*. Singapore National Printers, Singapore.

Singapore Department of Statistics (1993) *Yearbook of Statistics: Singapore*. Singapore National Printers, Singapore.

Strong, A. L. (1971) *Planned Urban Environments: Sweden, Finland, Israel, the Netherlands, France*. Johns Hopkins Press, Baltimore.

Taiwan Research Institute (1996), Web site: http://www.taiwaninformation.org/almanac/backgr2.html.

Thomas, D. (1983) *Flexibility and Commitment in Planning: a Comparative Study of Local Planning and Development in the Netherlands and England*. Nijhoff, The Hague.

Tien, Hung-mao (1989) *The Great Transition: Political and Social Change in the Republic of China*. Hoover Institution Press, Stanford University, Calif.

Urban Redevelopment Authority (1991) *Living the Next Lap: Towards a Tropical City of Excellence*. Urban Redevelopment Authority, Singapore.

Urban Redevelopment Authority (1993) *Master Plan Written Statement 1993, The Planning Act* (Cap 232, Revised Ed. 1990). Urban Redevelopment Authority, Singapore.

Urban Redevelopment Authority (1994) *Exhibition on Master Plan Proposals, Singapore River Planning Area*, pamphlet printed by Urban Redevelopment Authority.

Urban Redevelopment Authority (1995) *Changing the Face of Singapore through the URA Sale of Sites*. Urban Redevelopment Authority, Singapore.

Urban Redevelopment Authority (1996) WWW site: http://www.ura.gov.sg/about/text/lib-ura.html.

Wade, R. (1990) *Governing the Market: Economic Theory and the Role of Government in East Asian Industrialization.* Princeton University Press, Princeton, NJ.
White Paper (1991) *Shared Values.* Singapore National Printers, Singapore.
White, P. M. (1979) Urban planning in Britain and the Soviet Union: A comparative analysis. *Research Memorandum 70.* Joint Centre for Regional, Urban and Local Government Studies, University of Birmingham, Birmingham, U.K.
Wolman, H. (1985) National-urban relations in foreign federal systems: lessons for the United States. In *Urban Policy in a Changing Federal System,* Proceedings of a Symposium, ed. Warren, C. R. National Academy Press, Washington.
World Bank (1993) *The East Asian Miracle: Economic Growth and Public Policy.* Policy Research Report. The World Bank, Washington D.C.
Xin, Wan Jiao (1991) *Urban and Regional Planning.* Research Institute on Land Administration (in Chinese).
Yen, C. S. (1990) *Consequences of Socioeconomic Development in Taiwan, Singapore and South Korea: Democracy or a Continuation of Authoritarianism?* UMI, Ann Arbor, Michigan.
Yeung, Y. M. (ed.) (1993) *Pacific Asia in the 21st Century.* The Chinese University Press, Hong Kong.

MIX
Papier | Fördert
gute Waldnutzung
FSC® C083411

Kolibri 360 GmbH
Ferdinand-Jühlke-Straße 7
99095 Erfurt, Deutschland
produktsicherheit@kolibri360.de